MTC-121

LINEAR ALGEBRA

[2 Credits]

Mathematics : Paper-I

For

First Year B.Sc. Computer Science : Semester-II
New Syllabus as per CBCS Pattern
from June 2019

M. D. Bhagat
Ex-Head Dept. of Mathematics
Tuljaram Chaturchand College,
Baramati (Dist. Pune)

R. S. Bhamare
Ex-Head of P.G. Dept. of Mathematics
New Arts, Science & Commerce College,
Ahmednagar

N. M. Phatangare
Assistant Prof. Dept. of Mathematics
Fergusson College, (Autonomous),
Pune 411 004

Dr. S. G. Purane
Associate Prof. & HOD of Mathematics
PES's Jamkhed Mahavidyalaya
Jamkhed, Ahmednagar 413 201

Dr. A. S. Khairnar
Assistant Professor & HOD of Mathematics
MES's Abasaheb Garware College,
Pune 411 004

S. D. Manjarekar
Assistant Professor
Department of Mathematics
MGV's Loknete Vyankatrao Hiray
Arts, Science & Commerce College, Nashik 422 003

N5058

F.Y.B.Sc. Linear Algebra Maths P-I Sem. II   **ISBN 978-93-89686-07-4**

First Edition : November 2019

©   : Authors

Published By:
NIRALI PRAKASHAN
Abhyudaya Pragati, 1312, Shivaji Nagar
Off J.M. Road, PUNE – 411005
Tel - (020) 25512336/37/39, Fax - (020) 25511379
Email : niralipune@pragationline.com

➢ **DISTRIBUTION CENTRES**
PUNE

Nirali Prakashan   : 119, Budhwar Peth, Jogeshwari Mandir Lane, Pune 411002,
(For orders within Pune)   Maharashtra, Tel : (020) 2445 2044, Mobile : 9657703145
                              Email : niralilocal@pragationline.com

Nirali Prakashan   : S. No. 28/27, Dhayari, Near Asian College Pune 411041
(For orders outside Pune)   Tel : (020) 24690204; Mobile : 9657703143
                              Email : bookorder@pragationline.com
MUMBAI

Nirali Prakashan   : 385, S.V.P. Road, Rasdhara Co-op. Hsg. Society Ltd.,
                              Girgaum, Mumbai 400004, Maharashtra;
                              Mobile : 9320129587 Tel : (022) 2385 6339 / 2386 9976,
                              Fax : (022) 2386 9976
                              Email : niralimumbai@pragationline.com

➢ **DISTRIBUTION BRANCHES**
JALGAON

Nirali Prakashan   : 34, V. V. Golani Market, Navi Peth, Jalgaon 425001,
                              Maharashtra, Tel : (0257) 222 0395, Mob : 94234 91860;
                              Email : niralijalgaon@pragationline.com
KOLHAPUR

Nirali Prakashan   : New Mahadvar Road, Kedar Plaza, 1st Floor Opp. IDBI Bank,
                              Kolhapur 416 012, Maharashtra. Mob : 9850046155;
                              Email : niralikolhapur@pragationline.com
NAGPUR

Nirali Prakashan   : Above Maratha Mandir, Shop No. 3, First Floor,
                              Rani Jhanshi Square, Sitabuldi, Nagpur 440012, Maharashtra
                              Tel : (0712) 254 7129;
                              Email : niralinagpur@pragationline.com
DELHI

Nirali Prakashan   : 4593/15, Basement, Agarwal Lane, Ansari Road, Daryaganj
                              Near Times of India Building, New Delhi 110002
                              Mob : 08505972553, Email : niralidelhi@pragationline.com
BENGALURU

Nirali Prakashan   : Maitri Ground Floor, Jaya Apartments, No. 99, 6th Cross,
                              6th Main, Malleswaram, Bengaluru 560003, Karnataka;
                              Mob : 9449043034
                              Email: niralibangalore@pragationline.com
Other Branches : Hyderabad, Chennai

niralipune@pragationline.com | www.pragationline.com

Also find us on www.facebook.com/niralibooks

Preface ...

We have great pleasure in presenting this text book on **LINEAR ALGEBRA** to the students of F.Y.B.Sc. and B.A. Semester - II, Mathematics Paper - II. This book is written strictly according to the new revised syllabus of Savitribai Phule Pune University to be implemented from June 2019.

We have taken utmost care to present the matter systematically and with proper flow of mathematical concepts. We begin the Chapter by Introduction and at the end the Summary of the Chapter is provided. We have added one significant feature: **"Think Over It"** in this new **edition**. Here, we have posed questions of simple, difficult and intuitive type in nature. It is expected that the students should think over it and try to find the answers. This will assess the understanding of the knowledge of the Chapter.

The book contains good number of solved problems and the number of graded problems in the exercises.

We are thankful to **Shri Dineshbhai Furia, Shri Jignesh Furia,** Mrs. Anagha Medhekar (Proof Reading and Co-ordination), Mr. Ilyas Shaikh, Mrs. Anjali Muley (Fig. Drawing) and the staff of Nirali Prakashan for the great efforts that they have taken to publish the book in time.

We welcome the valuable suggestions from our colleagues' and readers for the improvement of the book.

PUNE   **AUTHORS**
NOVEMBER 2019

Syllabus ...

1. Vector Spaces   (10 Lectures)
- 1.1 Vector spaces and subspaces
- 1.2 Null spaces, column spaces and linear tranformations.
- 1.3 Linearly independent sets : Bases
- 1.4 Co-ordinate systems
- 1.5 The dimension of a vector space
- 1.6 Rank

2. Eigen Values and Eigen Vectors   (10 Lectures)
- 2.1 Eigen values and Eigen vectors
- 2.2 The characteristic equation
- 2.3 Diagonalization
- 2.4 Eigen vectors and Linear transformations

3. Orthogonality and Symmetric Matrices (10 Lectures)
- 3.1 Inner product, length and orthogonality
- 3.2 Orthogonal sets
- 3.3 Orthogonal Projections
- 3.4 Diagonalization of Symmetric Matrices
- 3.5 Quadratic forms

4. The Geometry of Vector Spaces   (6 Lectures)
- 4.1 Affine combinations
- 4.2 Affine independence
- 4.3 Convex combinations

☞ ☞ ☞

Contents ...

Chapter **1**...
Vector Spaces

Hermann Grassmann

In 1844 Hermann Grassmann published his "Theory of Extension" which included foundational new topics of what is today called Linear Algebra.

1.1 Introduction

In the (first semester) Matrix Algebra course, we have studied vectors in $\mathbb{R}^2$, $\mathbb{R}^3$ and $\mathbb{R}^n$. In $\mathbb{R}^n$, the vectors can be added and multiplied by scalars (real numbers) and these two operation on $\mathbb{R}^n$ satisfy axiomatic properties, which makes $\mathbb{R}^n$ significant and important tool in mathematics.

Mathematically, the input and output signals to an engineering system are functions. It is important in applications that these functions can be added and multiplied by scalars. These two operations on functions have algebraic properties that are completely analogous to the operations of adding vectors in $\mathbb{R}^n$ and multiplying by scalars. Because of this, the set of all inputs (functions) is called a **vector space.**

In this chapter, we are generalizing the properties and the characteristics of vectors in $\mathbb{R}^n$ to a more general context, and such sets (that having the same properties as $\mathbb{R}^n$) are called **real vector spaces**, as we will be using **only real numbers as scalars**. There is a more general concept of vector spaces, where the scalars may be complex numbers or may be elements of a field.

In this chapter, we shall study subspaces, linearly independent sets, bases, co-ordinate systems, the dimension of a vector spaces in more general context. We shall also return to Null spaces, Column spaces and Linear Transformations. The last section will be of rank of a matrix and that of linear transformation.

Vector Spaces - Definition and Examples :

The properties of addition of vectors in $\mathbb{R}^n$ and multiplying by scalars those we have seen earlier paves the way to the definition of a vector space.

Definition : A non-empty set V of objects, on which two operations are defined, called addition and multiplication by scalars (real numbers) is called a **vector space** if it satisfies the following ten axioms; the axioms must hold for all $\vec{u}$, $\vec{v}$ and $\vec{w}$ in V and for scalars α and β.

1. The addition of $\vec{u}$ and $\vec{v}$, denoted by $\vec{u} + \vec{v}$, is in V.

                                               (closureness of addition)

2. $\vec{u} + \vec{v} = \vec{v} + \vec{u}$ (+ is commutative)

3. $(\vec{u} + \vec{v}) + \vec{w} = \vec{u} + (\vec{v} + \vec{w})$ (+ is associative)

4. There is a zero vector $\vec{0}$ in V such that $\vec{u} + \vec{0} = \vec{u}$.

5. For each $\vec{u}$ in V, there is a vector $-\vec{u}$ such that $\vec{u} + (-\vec{u}) = \vec{0}$.

6. The scalar multiple of $\vec{u}$ by α, denoted by $\alpha\vec{u}$ is in V.

7. $\alpha(\vec{u} + \vec{v}) = \alpha\vec{u} + \alpha\vec{v}$.

8. $(\alpha + \beta)\vec{u} = \alpha\vec{u} + \beta\vec{u}$.

9. $(\alpha\beta)\vec{u} = \alpha(\beta\vec{u})$.

10. $1\vec{u} = \vec{u}$.

Examples of Vector Space :

1. For $n \geq 1$, the spaces $\mathbb{R}^n$, are the examples of vector space. The geometric institutions that we have seen in Semester - I for $\mathbb{R}^2$ and $\mathbb{R}^3$ are important to visualize many concepts in this chapter.

2. An expression $p(x) = a_0 + a_1x + \ldots + a_nx^n$ ... (i)

   is called a polynomial in x, where $a_0, a_1, \ldots, a_n$ are called coefficients and are real numbers, x is a real variable. The **degree** of a polynomial $p(x)$ is the highest power of x in (1) whose coefficient is non-zero. If $p(x) = a_0 \neq 0$, the degree of $p(x)$ is defined to be zero. If $p(x) \equiv 0$, that is, all the coefficients of $p(x)$ are zero, then $p(x)$ is called **zero polynomial** and **its degree is not defined**.

If $p(x)$ is given by (1) and if $q(x) = b_0 + b_1x + \ldots + b_nx^n$, then the sum $p + q$ is defined by

$$(p + q)(x) = p(x) + q(x)$$
$$= (a_0 + b_0) + (a_1 + b_1)x + \ldots + (a_n + b_n)x^n$$

And the scalar multiple αp is defined by

$$(\alpha p)(x) = \alpha p(x) = (\alpha a_0) + (\alpha a_1)x + \ldots (\alpha a_n)x^n$$

For $n \geq 0$, let P_n be the set of all polynomials of degree $\leq n$, that is,

$$P_n = \{a_0 + a_1x + \ldots + a_nx^n \mid a_0, a_1, \ldots, a_n \in \mathbb{R}, x \text{ is variable}\}$$

Under the addition and scalar multiplication as defined above, we see that $p + q$ and αp are polynomials of degree $\leq n$, hence are members of P_n. It is easy to see that these operations satisfy the axioms of a vector space. The zero polynomial acts as zero vector and $- p$ is defined to be

$$(-p)(x) = -p(x) = (-a_0) + (-a_1)x + \ldots + (-a_n)x^n$$

Thus, P_n is a vector space.

3. Let V be the set of all real valued functions defined on D, where D is any subset of $\mathbb{R}$. Thus, $V = \{f \mid f : D \to \mathbb{R},\ D \subseteq \mathbb{R}\}$

   We know :

   (i)   Addition of two functions f and g is defined as

   $$(f + g)(x) = f(x) + g(x),\ x \in D \text{ and}$$

   (ii) Scalar multiple αf as, $(\alpha f)(x) = \alpha f(x),\ x \in D.$

   (iii) Equality of two functions of f and g is defined as $f = g$, if and only if $f(x) = g(x),\ \forall\ x \in D.$

   (iv) Zero vector in V is a function f which is identically zero, that is, $f(x) = 0$, for all x in D.

   (v)   The negative of the function $f \in V$ is defined to be $- f$ as

   $$(-f)(x) = - f(x),\ x \in D.$$

   From (i) to (v) and the properties of the real numbers, we can verify that **V is a vector space.**

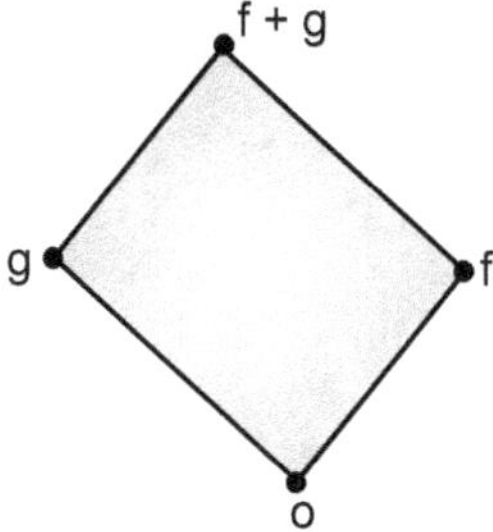

Fig. 1.1 : The Sum of the Two Vectors (Functions)

Note : If in example (2), $x \in D$, then P_n is set of real valued functions defined on D, which are called polynomial functions, and clearly $P_n \subset V$.

4. Let V be the set of all arrows (directed line segments) in three dimensional space, with (i) two arrows are regarded equal if they have the same length and point in the same direction. Define addition by the parallelogram rule, and for each $\vec{v}$ in V define $\alpha\vec{v}$ to be the length of $\vec{v}$, pointing in the direction as $\vec{v}$ if $\alpha \geq 0$, and otherwise pointing in the opposite direction. Then V is a vector space.

The definition of V, in this case is geometric, using length and direction. No xyz - co-ordinate system is involved. **An arrow of zero length** is a single point and represents zero vector. The negative of a vector $\vec{v}$ is $(-1)\,\vec{v}$ so the axioms 1, 4, 5, 6, 10 are obvious. The rest of the axioms can be verified geometrically, see the Fig. 1.2 and 1.3 below.

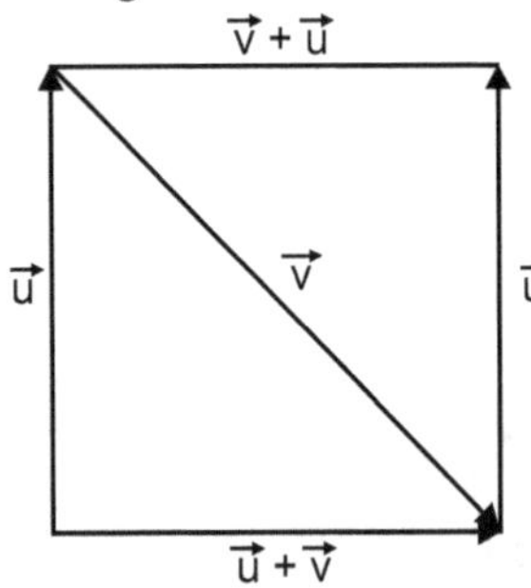

 Fig. 1.2   **Fig. 1.3**

Some Simple Consequences of Vector Axioms :

$\boxed{\textbf{Theorem 1}}$ Let V be a vector space, then :

(1) The zero vector in V is unique.

(2) The negative vector $-\vec{u}$ of $\vec{u}$ is unique.

(3) $0 \cdot \vec{u} = \vec{0}$, for all $\vec{u}$ in V.

(4) $\alpha \cdot \vec{0} = \vec{0}$ for any scalar.

(5) $-\vec{u} = (-1)\,\vec{u}$, for any $\vec{u} \in V$.

(6) $\alpha\vec{u} = \vec{0}$ if and only if either $\alpha = 0$ or $\vec{u} = \vec{0}$.

Proof : (1) Suppose there are two zero vectors say $\vec{0}$ and $\vec{0}'$, then

$$\vec{0} = \vec{0} + \vec{0}' \qquad\qquad \because \vec{0}' \text{ is zero vector}$$

$$= \vec{0}' \qquad\qquad \because \vec{0} \text{ is a zero vector}$$

$$\therefore \qquad \vec{0} = \vec{0}'$$

This proves that there is only one zero vector in V. ∎

(2) Suppose there are two negatives of $\vec{u}$ say $\vec{u}_1$ and $\vec{u}_2$, then by axiom (5).

$$\vec{u} + \vec{u}_1 = \vec{0} \text{ and } \vec{u} + \vec{u}_2 = \vec{0}.$$

Now, $\qquad \vec{u}_1 = \vec{0} + \vec{u}_1 \qquad\qquad$ By axiom (4)

$$= (\vec{u} + \vec{u}_2) + \vec{u}_1 \qquad\qquad \because \vec{u} + \vec{u}_2 = \vec{0}$$

$$= (\vec{u}_2 + \vec{u}) + \vec{u}_1 \qquad\qquad \text{By axiom (2)}$$

$$= \vec{u}_2 + (\vec{u} + \vec{u}_1) \qquad\qquad \text{By axiom (3)}$$

$$= \vec{u}_2 + \vec{0} \qquad\qquad \because \vec{u} + \vec{u}_1 = \vec{0}$$

$$= \vec{u}_2 \qquad\qquad \text{By axiom (4)}$$

$$\therefore \qquad \vec{u}_1 = \vec{u}_2$$

This shows that the negative of the vector $\vec{u}$ is unique, we denote it by $-\vec{u}$.

(3) Since, we know $0 + 0 = 0$ in $\mathbb{R}$.

$$\therefore \ (0 + 0) \vec{u} = 0 \cdot \vec{u} \text{ for any } \vec{u} \text{ in V.}$$

$$0 \cdot \vec{u} + 0 \cdot \vec{u} = 0 \cdot \vec{u} \qquad\qquad \text{Using axiom (8)}$$

Adding negative of $0\vec{u}$ on both sides we obtain,

$$0 \cdot \vec{u} = \vec{0}$$

(4) Since $\vec{0}$ is a zero vector in V,

then $\qquad \vec{0} + \vec{0} = \vec{0}$

So for any scalar α

$$\alpha(\vec{0} + \vec{0}) = \alpha \cdot \vec{0}$$

$$\alpha\,\vec{0} + \alpha\vec{0} = \alpha\vec{0} \qquad \text{By axiom (7)}$$

Again adding on both sides the negative of $\alpha \cdot \vec{0}$, we obtain

$$\alpha \cdot \vec{0} = \vec{0}$$

(5) For any $\vec{u}$ in V, we have

$$\vec{u} + (-1)\,\vec{u} = 1 \cdot \vec{u} + (-1)\,\vec{u} \qquad \text{By axiom (10)}$$

$$= (1 + (-1))\,\vec{u} \qquad \text{By axiom (8)}$$

$$= 0 \cdot \vec{u} \qquad \because 1 + (-1) = 0 \text{ in } \mathbb{R}$$

$$= \vec{0}$$

Showing that $(-1)\,\vec{u}$ is negative of $\vec{u}$ by axiom (5), hence,

$$-\,\vec{u} = (-1)\,\vec{u}.$$

(6) If $\alpha = 0$, then by (3) above $0 \cdot \vec{u} = \vec{0}$

and if $\vec{u} = \vec{0}$, then by (4) above $\alpha \cdot \vec{0} = \vec{0}$.

So, if either $\alpha = 0$ or $\vec{u} = \vec{0}$, then $\alpha\vec{u} = \vec{0}$.

Conversely, suppose $\alpha\vec{u} = \vec{0}$. If $\alpha \neq 0$, then we have,

$$\alpha\vec{u} = \vec{0}$$

$$\Rightarrow \qquad \frac{1}{\alpha}(\alpha\vec{u}) = \frac{1}{\alpha}\vec{0} \qquad \because \alpha \neq 0$$

$$\Rightarrow \qquad \left(\frac{1}{\alpha} \cdot \alpha\right)\vec{u} = \vec{0} \qquad \text{By (3) above and axiom (9)}$$

$$\Rightarrow \qquad 1\,\vec{u} = \vec{0}$$

$$\Rightarrow \qquad \vec{u} = \vec{0} \qquad \text{By axiom (10)}$$

Examples of Vector Spaces Continued :

5. Let $V = M_{m \times n}(\mathbb{R})$ be the set of all $m \times n$ matrices with real entries. Then for $A = [a_{ij}]_{m \times n}$, $B = [b_{ij}]_{m \times n}$, we have defined,

$$A + B = [a_{ij} + b_{ij}]_{m \times n} \text{ and for any scalar } \alpha,$$

$$\alpha A = [\alpha a_{ij}]$$

   Then we have seen the properties of matrix addition and scalar multiplication, it is clear that V is a vector space.

6. Let $W = C(D)$ be the set of all continuous real valued functions defined on D. Then with addition and scalar multiplication as defined in example (3), one can verify using properties of continuous functions that W is a vector space.

   Note : W is clearly a subset of V in example (3).

7. Let S be the space of all doubly infinite sequences, usually written in row :

$$\{x_n\} = \{\ldots x_{-2}, \; x_{-1}, \; x_0, \; x_1, \; x_2, \; \ldots\}$$

   if $\{y_n\}$ is another element of S, then the sum $\{x_n\} + \{y_n\}$ is a sequence $\{x_n + y_n\}$ formed by adding the corresponding terms of $\{x_n\}$ and $\{y_n\}$. The scalar multiple $\alpha\{x_n\}$ is the sequence $\{\alpha x_n\}$. The vector space axioms are verified in the same way as for $\mathbb{R}^n$.

1.2 Subspaces

A vector space which is a subset of a larger vector space is called a subspace of a larger vector space.

Definition : A subspace of a vector space V is a subset W of V that has three properties :

(a) The zero vector of V is in W.

(b) W is closed under vector addition, that is, for any $\vec{u}, \vec{v}$ in W, $\vec{u} + \vec{v} \in W$.

(c) W is closed under multiplication by scalars. That is, for each $\vec{u}$ in W and for each scalar α, the vector $\alpha\vec{u}$ is in W.

For example,

1. The set consisting of only the zero vector in a vector space V is a subspace of V, called the **zero subspace** and is denoted by $\{\vec{0}\}$.

2. Let P denote the set of all polynomials with real coefficients then under addition and scalar multiplication defined as in example (2) of the vector spaces, we can verify that p is a vector space. Clearly, P_n the vector space of polynomials of degree $\leq n$ is a subset of P, hence P_n is a subspace of P.

3. Consider P and P_n as functions of real variable, that is, $x \in \mathbb{R}$, then P and P_n are subsets of the function space V of all real valued functions defined on $\mathbb{R}$ as a domain. Then P and P_n are clearly subspaces of V.

4. *A line in $\mathbb{R}^2$, not passing through the origin is not a* **subspace** *of $\mathbb{R}^2$. A plane in $\mathbb{R}^3$ not passing through the origin is not a subspace of $\mathbb{R}^3$, as the zero vector in both cases does not belong to the line or the plane.*

Example 1.1 : *Show that a subspace W of a vector space V is a vector space.*

Solution : The conditions (a), (b) and (c) for a subspace are the axioms (1), (4) and (6) for vector space. The axioms (2), (3) and (7) to (10) are automatically true in W because they apply to all elements of V and hence for elements in W, as $W \subseteq V$. The axiom (5) also holds for W, because if $\vec{u}$ is in W, then $(-1)\,\vec{u} \in W$, by (c).

But we have to prove that $-\,\vec{u} = (-1)\,\vec{u}$, so $-\,\vec{u} \in W$. Thus, W satisfies all the vector space axioms, hence is a vector space itself.

Note : Conversely, every vector space is a subspace of itself (or possibly that of larger vector space).

Example 1.2 : *Let $W = \{a_0 + a_1x + a_2x^2 \,/\, a_0, a_1, a_2 \in \mathbb{R}\}$. Show that W is a subspace of P_n, $n \geq 2$.*

Solution : First of all W is subset of P_n, $n \geq 2$, as any element $a_0 + a_1x + a_2x^2 \in W$, the third term onwards the coefficients can be supposed to be zero. Since, $0 = 0 + 0.x + 0.x^2$, the zero polynomial belongs to W. For any two members $P(x) = a_0 + a_1x + a_2x^2$ and $q(x) = b_0 + b_1x + b_2x^2$, with $a_0, a_1, a_2; b_0, b_1, b_2$ in $\mathbb{R}$ in W, we have,

$$p(x) + q(x) = (a_0 + b_0) + (a_1 + b_1)\,x + (a_2 + b_2)\,x^2$$

clearly $p(x) + q(x) \in W$.

Also, for any scalar (real number) α,

$$\alpha p(x) = \alpha(a_0 + a_1x + a_2x^2)$$
$$= (\alpha a_0) + (\alpha a_1)\,x + (\alpha a_2)\,x^2$$

Hence, $\alpha p(x)$ belongs to W.

Thus, W satisfies all the three conditions for the subspace hence W is a subspace of P_n.

Example 1.3 : *The vector space $\mathbb{R}^2$ is not a **subspace** of $\mathbb{R}^3$, because $\mathbb{R}^2$ is not even a subset of $\mathbb{R}^3$, as vectors in $\mathbb{R}^3$ have three entries whereas the vectors in $\mathbb{R}^2$ have only two entries.*

However, the set $W = \left\{ \begin{bmatrix} s \\ t \\ 0 \end{bmatrix} : s, t \text{ are real} \right\}$ is a subset of $\mathbb{R}^3$, that "looks" and "acts" like $\mathbb{R}^2$, although it is a logically distinct from $\mathbb{R}^2$. Show that W is a subspace of $\mathbb{R}^3$.

Solution : Since, s, t $\in \mathbb{R}$, $\begin{bmatrix} 0 \\ 0 \\ 0 \end{bmatrix}$ belongs to W.

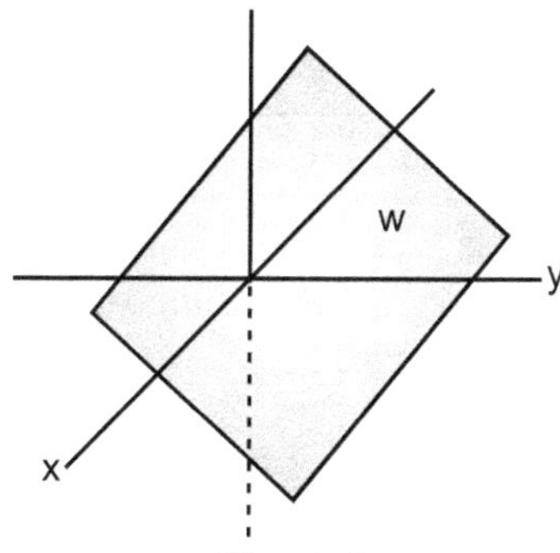

Fig. 1.4

That is the zero vector is in W, so first condition is satisfied. It is easy to see that W is closed under vector addition and scalar multiplication because these operations on vectors in W produce vectors in W. Thus, W is a subspace of $\mathbb{R}^3$. See the Fig. 1.4.

Note : A line through origin in $\mathbb{R}^2$ is a subspace of $\mathbb{R}^2$.

Example 1.4 : *Let W be set in the xy-plane given by*

$$W = \left\{ \begin{bmatrix} x \\ y \end{bmatrix} : xy \ge 0 \right\}.$$

(W is a union of the first and third quadrants in xy-plane ($\mathbb{R}^2$))

(a) If $\vec{u}$ is in W and α is any scalar, is $\alpha\vec{u}$ in W ? Why ?

(b) Find specific vectors $\vec{u}$ and $\vec{v}$ in W such that $\vec{u} + \vec{v}$ is not in W.

Solution : (a) For any $\vec{u} = \begin{bmatrix} x \\ y \end{bmatrix}$ in W, we have $xy \ge 0$, if α is any scalar, then $\alpha\vec{u} = \alpha \begin{bmatrix} x \\ y \end{bmatrix} = \begin{bmatrix} \alpha x \\ \alpha y \end{bmatrix}$ and $(\alpha x) \times (\alpha y) = \alpha^2 xy \ge 0$.

Since $xy \geq 0$, and α is real number

so $\alpha^2 \geq 0$, therefore $\alpha \vec{u} \in W$.

That is, W is closed under scalar multiplication.

(b)   Let $\vec{u} = \begin{bmatrix} 2 \\ 3 \end{bmatrix}$ and $\vec{v} = \begin{bmatrix} -5 \\ -1 \end{bmatrix}$ in W as $2 \times 3 = 6 > 0$ and $(-5)(-1) > 0$. But

$$\vec{u} + \vec{v} = \begin{bmatrix} 2 \\ 3 \end{bmatrix} + \begin{bmatrix} -5 \\ -1 \end{bmatrix} = \begin{bmatrix} 2-5 \\ 3-1 \end{bmatrix} = \begin{bmatrix} -3 \\ 2 \end{bmatrix}.$$

Since, $(-3) \times 2 = -6 < 0$, hence, $\begin{bmatrix} -3 \\ 2 \end{bmatrix}$ is not in W or that is $\vec{u} + \vec{v}$ is not in W. Thus, W is not closed under vector addition, hence W is not a subspace of the xy-plane.

A Subspace Spanned by a Set :

Let $\vec{u}_1, \vec{u}_2, \ldots \vec{u}_k$ be vectors in a vector space V. The linear combination of these vectors is any sum of scalar multiples of the vectors $\vec{u}_1, \ldots, \vec{u}_k$. That is, $\alpha_1 \vec{u}_1 + \alpha_2 \vec{u}_2 + \ldots + \alpha_k \vec{u}_k$, where $\alpha_1, \ldots, \alpha_k$ are scalars, is a linear combination of the vectors $\vec{u}_1, \ldots, \vec{u}_k$.

Span $\{\vec{u}_1, \vec{u}_2, \ldots, \vec{u}_k\}$ is the set of all vectors in V that can be written as linear combination of $\vec{u}_1, \ldots, \vec{u}_k$. That is,

Span $\{\vec{u}_1, \ldots, \vec{u}_k\} = \{\alpha_1 \vec{u}_1 + \ldots + \alpha_k \vec{u}_k \, / \, \alpha_1, \alpha_2, \ldots, \alpha_k$ are scalars$\}$

Theorem 2 If $\vec{u}_1, \vec{u}_2, \ldots, \vec{u}_k$ are vectors in a vector space V, then

Span $\{\vec{u}_1, \vec{u}_2, \ldots, \vec{u}_k\}$ is a subspace of V.

Proof : Clearly $\vec{0} = 0.\vec{u}_1 + 0.\vec{u}_2 + \ldots + 0\vec{u}_k$ is in Span $\{\vec{u}_1, \vec{u}_2, \ldots, \vec{u}_k\}$.

Now if $\vec{u}$ and $\vec{v}$ are any vectors in Span $\{\vec{u}_1, \vec{u}_2, \ldots, \vec{u}_k\}$, then we have

$$\vec{u} = \alpha_1 \vec{u}_1 + \ldots + \alpha_k \vec{u}_k, \quad \alpha_1, \alpha_2, \ldots, \alpha_k \text{ scalars and}$$

$$\vec{v} = \beta_1 \vec{u}_1 + \ldots + \beta_k \vec{u}_k, \quad \beta_1, \beta_2, \ldots, \beta_k \text{ scalars.}$$

Now, $\vec{u} + \vec{v} = (\alpha_1 \vec{u}_1 + \ldots + \alpha_k \vec{u}_k) + (\beta_1 \vec{u}_1 + \ldots + \beta_k \vec{u}_k)$

Using vector space axioms, we can write $\vec{u} + \vec{v}$ as :

$$\vec{u} + \vec{v} = (\alpha_1 + \beta_1) \vec{u}_1 + \ldots + (\alpha_k + \beta_k) \vec{u}_k,$$

so clearly $\vec{u} + \vec{v}$ is in Span $\{\vec{u}_1, ..., \vec{u}_k\}$.

Also, for any scalar a, we have using vector space axioms,

$$a\vec{u} = a(\alpha_1\vec{u}_1 + ... + \alpha_k\vec{u}_k)$$

$$= (a\alpha_1)\,\vec{u}_1 + ... + (a\alpha_k)\,\vec{u}_k$$

and this shows that $a\vec{u} \in$ Span $\{\vec{u}_1, ..., \vec{u}_k\}$.

Thus, $\{\vec{u}_1, \vec{u}_2, ... , \vec{u}_k\}$ satisfies all three conditions of a subspace, therefore is a subspace of V.   ■

Remark :

1. Given two vectors $\vec{u}_1$ and $\vec{u}_2$ in a vector space V, the set $W = $ Span $\{\vec{u}_1, \vec{u}_2\}$ is a subspace of V, which is clear from theorem 1 as particular case with $k = 2$.

2. We call Span$\{\vec{u}_1, ..., \vec{u}_k\}$ the **subspace spanned** (or **generated**) by $\{\vec{u}_1, ..., \vec{u}_k\}$.

    Given any subspace W of V, a **spanning** (or **generating**) set for W is the set $\{\vec{u}_1, ..., \vec{u}_k\}$ in W such that

    $$W = \text{Span } \{\vec{u}_1, ..., \vec{u}_k\}.$$

Example 1.9 : *Let W be the set of all vectors of the form* $\begin{bmatrix} s + 3t \\ s - t \\ 2s - t \\ 4t \end{bmatrix}$.

Show that W is a subspace of $\mathbb{R}^4$.

Solution : We write vectors in W as column vectors. Then the arbitrary vector in W has the form.

$$\begin{bmatrix} s + 3t \\ s - t \\ 2s - t \\ 4t \end{bmatrix} = s\begin{bmatrix} 1 \\ 1 \\ 2 \\ 0 \end{bmatrix} + t\begin{bmatrix} 3 \\ -1 \\ -1 \\ 4 \end{bmatrix}$$

$$\qquad\qquad\qquad \underset{\vec{u}_1}{\uparrow} \qquad \underset{\vec{u}_2}{\uparrow}$$

This shows that $W = $ Span $\{\vec{u}_1, \vec{u}_2\}$, where $\vec{u}_1$ and $\vec{u}_2$ are vectors indicated above. Thus, W is a subspace of $\mathbb{R}^4$ by theorem 1.

Example 1.10 : *Let $V = M_{2 \times 2}$ ($\mathbb{R}$) be the vector space of all 2×2 matrices with real entries; which of the following subsets of V are subspaces :*

(a) $W_1 = \left\{ \begin{bmatrix} a & b \\ 0 & d \end{bmatrix} \,\middle|\, a, b, d \in \mathbb{R} \right\}$

(b) $W_2 = \left\{ \begin{bmatrix} a & b \\ c & d \end{bmatrix} \,\middle|\, a, b, c \in \mathbb{Q} \right\}$

(c) $W_3 = \left\{ \begin{bmatrix} a & b \\ 0 & 0 \end{bmatrix} \,\middle|\, a, b \in \mathbb{Z} \right\}$

(d) $W_4 = \left\{ \begin{bmatrix} a & 0 \\ 0 & d \end{bmatrix} \,\middle|\, a, d \in \mathbb{R} \text{ and } ad \geq 0 \right\}$

(e) $W_5 = \left\{ \begin{bmatrix} a & 2 \\ c & d \end{bmatrix} \,\middle|\, a, c, d \in \mathbb{R} \right\}$

Solution : (a) For any two matrices $\begin{bmatrix} a & b \\ 0 & d \end{bmatrix}$ and $\begin{bmatrix} a_1 & b_1 \\ 0 & d_1 \end{bmatrix}$ in W_1.

We see that $\begin{bmatrix} a & b \\ 0 & d \end{bmatrix} + \begin{bmatrix} a_1 & b_1 \\ 0 & d_1 \end{bmatrix} = \begin{bmatrix} a + a_1 & b + b_1 \\ 0 & d + d_1 \end{bmatrix} \in W_1$

So W_1 is closed under addition.

Also for any α (scalar) in $\mathbb{R}$,

$$\alpha \begin{bmatrix} a & b \\ 0 & d \end{bmatrix} = \begin{bmatrix} \alpha a & \alpha b \\ 0 & \alpha d \end{bmatrix} \in W_1, \text{ hence } W_1 \text{ is closed under scalar}$$

multiplication.

It is clear that the zero matrix $\begin{bmatrix} 0 & 0 \\ 0 & 0 \end{bmatrix} \in W_1$.

$\therefore$ W_1 is a subspace of V.

(b)   Observe that for $\begin{bmatrix} \frac{2}{3} & 1 \\ -5 & -3 \end{bmatrix} \in W_2$ and if we take $\alpha = \sqrt{2}$

(scalar) in $\mathbb{R}$, $\alpha \begin{bmatrix} \frac{2}{3} & 1 \\ 5 & -3 \end{bmatrix} = \begin{bmatrix} \sqrt{2} \times \frac{2}{3} & 1 \times \sqrt{2} \\ 5 \times \sqrt{2} & (-3)\sqrt{2} \end{bmatrix}$.

We see that no entry in this matrix is rational so $\alpha \begin{bmatrix} \frac{2}{3} & 1 \\ 5 & -3 \end{bmatrix} \notin W_2$, that is, W_2 is not closed under scalar multiplication hence, is not a subspace.

(c) Similar to (b) above W_3 is not closed under scalar multiplication, hence is not a subspace.

(d) In this case observe that $\begin{bmatrix} 2 & 0 \\ 0 & 5 \end{bmatrix} \in W_4$, as $2 \times 5 = 10 > 0$, but for choice of $\alpha = -1$ (scalar) $\alpha \begin{bmatrix} 2 & 0 \\ 0 & 5 \end{bmatrix} = (-1) \begin{bmatrix} 2 & 0 \\ 0 & 5 \end{bmatrix} = \begin{bmatrix} -2 & 0 \\ 0 & -5 \end{bmatrix}$ and $(-2) \times (-5) = 10 > 0$, hence W_4 is closed under scalar multiplication.

But $\begin{bmatrix} 2 & 0 \\ 0 & 5 \end{bmatrix}$ and $\begin{bmatrix} -7 & 0 \\ 0 & -2 \end{bmatrix}$ both belong to W_4, as $2 \times 5 = 10 > 0$ and $(-7) \times (-2) = 14 > 0$.

But $\begin{bmatrix} 2 & 0 \\ 0 & 5 \end{bmatrix} + \begin{bmatrix} -7 & 0 \\ 0 & -2 \end{bmatrix} = \begin{bmatrix} -5 & 0 \\ 0 & 3 \end{bmatrix} \notin W_4$ hence, W_4 is not closed under addition, hence is not subspace.

(e) Here clearly $\begin{bmatrix} 0 & 0 \\ 0 & 0 \end{bmatrix}$ zero matrix is not in W_5, hence is not a subspace.

Example 1.11 : *Determine if the given set is a subspace of P_n, for appropriate value of n. Justify your answer.*

(a) *All polynomials of the form $p(x) = a_0 + a_1x + a_2x^2$, a_0, a_1, $a_2 \in \mathbb{R}$*

(b) *All polynomials of the form $p(x) = a_0 + 2x^2$, $a_0 \in \mathbb{R}$.*

(c) *All polynomials of degree at most 5, with integer coefficients.*

(d) *All polynomials in P_n with constant term zero.*

Solution : (a) Since, each polynomial in this set is generated by 1, x and x^2 hence by theorem (1) is a subspace.

(b) The zero polynomial cannot be in this set so it is not a subspace.

(c) This set is not closed under scalar multiplication, hence is not a subspace.

(d) This set a subspace because is generated by the set x, x^2, ..., (by theorem 1).

1.3 Null Space, Column Spaces, and Linear Transformations

The concepts Null space, Column space and linear transformations are already known which we used as subspaces of $\mathbb{R}^n$ (first two). In this section, we study these concepts to some more general context and especially subspace of general vector space. Here, these concepts will be seen in a way that it will describe the distinction between solutions to a system of homogeneous equations (**null space**) and the set of linear combination of certain specified vectors in a vector space (**column space**).

Definition : The **null space** of an m × n matrix A, written as Nul A, is the set of all solutions to the homogeneous equation $A\vec{X} = \vec{0}$.

Thus, Nul A = $\{\vec{x} \in \mathbb{R}^n \mid A\vec{x} = \vec{0}\}$.

Remark : As the map : $T : \mathbb{R}^n \to \mathbb{R}^m$ which maps each $\vec{x} \in \mathbb{R}^n$ to $A\vec{x}$ is a linear transformation, so in other words, Nul A is the set of all $\vec{x}$ in $\mathbb{R}^n$ that are mapped to the zero vector of $\mathbb{R}^m$ under T. See the following Fig. 1.5.

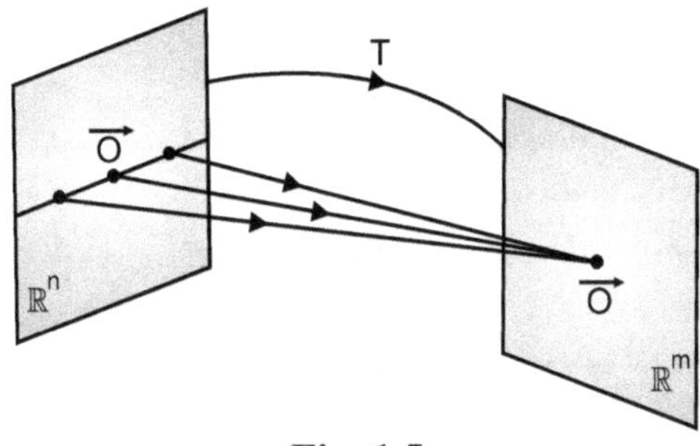

Fig. 1.5

Theorem 3 The null space of an m × n matrix A is a subspace of $\mathbb{R}^n$. In other words, the set of all solutions to a system $A\vec{x} = \vec{0}$ of m homogeneous linear equations in n unknowns is a subspace of $\mathbb{R}^n$.

Proof : Clearly Nul A is a subset of $\mathbb{R}^n$ because A has n columns. We have to show that Nul A satisfies the three properties of a subspace.

Since, $A\vec{0} = \vec{0}$, $\vec{0}$ is in Nul A.

For any $\vec{x}$ and $\vec{y}$ in $\mathbb{R}^n$, for $\vec{x} + \vec{y}$ is in Nul A, we have to show that

$$A(\vec{x} + \vec{y}) = \vec{0}$$

We have, $\quad A(\vec{x} + \vec{y}) = A\vec{x} + A\vec{y}$ $\qquad$ By matrix multiplication

$$= \vec{0} + \vec{0} \qquad \because \vec{x} + \vec{y} \text{ are in Nul A.}$$

$$= \vec{0}$$

Thus, $\vec{x} + \vec{y}$ is in Nul A.

Also, for any scalar α,

$$A(\alpha\vec{x}) = \alpha(A\vec{x}) \qquad \text{By property of matrix multiplication}$$

$$= \alpha\vec{0}$$

$$= \vec{0}$$

$\therefore \quad \alpha\vec{x}$ is a Nul A.

Thus, Nul A satisfies all three conditions, hence Nul A is a subspace of $\mathbb{R}^n$. $\qquad\blacksquare$

Example 1.12 : *Determine if* $\vec{w} = \begin{bmatrix} 5 \\ -3 \\ 2 \end{bmatrix}$ *is in Nul A, where,*

$$A = \begin{bmatrix} 5 & 21 & 19 \\ 13 & 23 & 2 \\ 8 & 14 & 1 \end{bmatrix}$$

Solution : We have to see that $\vec{w}$ satisfies $A\vec{x} = \vec{0}$, when $\vec{x}$ is replaced by $\vec{w}$.

$$A\vec{w} = \begin{bmatrix} 5 & 21 & 19 \\ 13 & 23 & 2 \\ 8 & 14 & 1 \end{bmatrix}\begin{bmatrix} 5 \\ -3 \\ 2 \end{bmatrix}$$

$$= \begin{bmatrix} 5 \times 5 + 21 \times (-3) + 19 \times 2 \\ 13 \times 5 + 23 \times (-3) + 2 \times 2 \\ 8 \times 5 + 14 \times (-3) + 1 \times 2 \end{bmatrix}$$

$$= \begin{bmatrix} 25 - 63 + 38 \\ 65 - 69 + 4 \\ 40 - 42 + 2 \end{bmatrix}$$

$$= \begin{bmatrix} 0 \\ 0 \\ 0 \end{bmatrix} = \vec{0}$$

So, $\vec{w}$ is in Nul A.

Example 1.13 : *Show that the set*

$$W = \left\{ \begin{bmatrix} a \\ b \\ c \\ d \end{bmatrix} : a - 2b = 4c,\ 2a = c + 3d \right\} \text{ is a vector space.}$$

Solution : The conditions $a - 2b = 4c$ and $2a = c + 2d$ can be written as

$$a - 2b - 4c = 0$$
$$2a - c - 3d = 0$$

This is a homogeneous system with two equations and in four variables hence the set W is a subspace by theorem 2, and is a vector space.

Note : There is no relation of Nul A with entries/columns of A, so we say that Nul A is defined implicitly. The question is how to describe vectors in Nul A or how to find the spanning set of Nul A. For this we illustrate this by the following example.

An Explicit Description of Nul A :

Example 1.14 : *Find the spanning set of the null space of A, where,*

$$A = \begin{bmatrix} 1 & -4 & -2 & 0 & 3 & -5 \\ 0 & 0 & 1 & 0 & 0 & -1 \\ 0 & 0 & 0 & 0 & 1 & -4 \\ 0 & 0 & 0 & 0 & 0 & 0 \end{bmatrix}.$$

Solution : We need to find the general solution of $A\vec{x} = \vec{0}$ in terms of free variables. For this, in the present case, the matrix A is already in row echelon form (if it is not one has to reduce it to row-echelon form and then next into reduced row echelon form), we convert it into reduced row echelon form by $R_1 + 2R_2$ and then $R_1 - 3R_3$. We have,

$$A \sim \begin{bmatrix} 1 & -4 & 0 & 0 & 0 & 6 \\ 0 & 0 & 1 & 0 & 0 & -1 \\ 0 & 0 & 0 & 0 & 1 & -4 \\ 0 & 0 & 0 & 0 & 0 & 0 \end{bmatrix}$$

The corresponding system of equations to $A\vec{x} = \vec{0}$ can be written as

$$x_1 - 4x_2 + 6x_6 = 0$$
$$x_3 - x_6 = 0$$
$$x_5 - 4x_6 = 0$$
$$0 = 0$$

Then the general solution is

$$x_1 = 4x_2 + 6x_6$$

$$x_3 = x_6$$

$$x_5 = 4x_6$$

With x_1, x_3 and x_5 are basic variables and x_2, x_4 and x_6 are free variables.

We express this general solution as a linear combination of vectors where the weights are free variables.

$$\begin{bmatrix} x_1 \\ x_2 \\ x_3 \\ x_4 \\ x_5 \\ x_6 \end{bmatrix} = \begin{bmatrix} 4x_2 + 6x_6 \\ x_2 \\ x_6 \\ x_4 \\ 4x_6 \\ x_6 \end{bmatrix} = x_2 \begin{bmatrix} 4 \\ 1 \\ 0 \\ 0 \\ 0 \\ 0 \end{bmatrix} + x_4 \begin{bmatrix} 0 \\ 0 \\ 0 \\ 1 \\ 0 \\ 0 \end{bmatrix} + x_6 \begin{bmatrix} 6 \\ 0 \\ 1 \\ 0 \\ 4 \\ 1 \end{bmatrix}$$

$$\qquad\qquad\qquad \underset{\vec{u}}{\uparrow} \qquad\quad \underset{\vec{v}}{\uparrow} \qquad\quad \underset{\vec{w}}{\uparrow}$$

$$= x_2 \vec{u} + x_4 \vec{v} + x_6 \vec{w}$$

Every linear combination of $\vec{u}$, $\vec{v}$, $\vec{w}$ is an element of Nul A.

Thus, $\{\vec{u}, \vec{v}, \vec{w}\}$ is a spanning set of Nul A.

Note : From above example, we should note following two points that apply to all problems of this type :

1. The spanning set obtained in the example is linearly independent because the free variables are the weights of the spanning vectors. Because

$$x_2 \vec{u} + x_4 \vec{v} + x_6 \vec{w} = \vec{0} \text{ only if } x_2 = 0 = x_4 = x_6.$$

2. When a null space contains a non-zero vector, the number of vectors in the spanning set for Nul A equals the number of free variables in solution of $A\vec{X} = \vec{0}$.

The Column Space of a Matrix :

Definition : The **column space** of an $m \times n$ matrix A, denoted by Col A, is the set of all linear combinations of the columns of A.

If $A = [\vec{a_1}, \ldots, \vec{a_n}]$, then Col A = Span $\{\vec{a_1}, \ldots, \vec{a_n}\}$.

Theorem 4 The column space of an $m \times n$ matrix A is a subspace of $\mathbb{R}^m$.

Proof : Since the column space of a matrix A is the spanning set of the columns of A and spanning set is a subspace. Since the columns of A are vectors in $\mathbb{R}^m$, Col A is a subspace of $\mathbb{R}^m$.

Note :

1. A typical vector in Col A can be written as $A\vec{x}$ for some $\vec{x}$, because the notation $A\vec{x}$ stands for linear combination of the columns of A. Thus,

$$\text{Col } A = \{ \vec{b} \in \mathbb{R}^m \mid \vec{b} = A\vec{x}, \text{ for some } \vec{x} \text{ in } \mathbb{R}^n \}.$$

2. Columns A is the **range set** of the **linear transformation** $\vec{x} \to A\vec{x}$.   ∎

Example 1.15 : *Find a matrix A such that the set*

$$W = \left\{ \begin{bmatrix} 2s + 3t \\ r + s - 2t \\ 4r + s \\ 3r - s - t \end{bmatrix} : r, s, t \text{ real numbers} \right\} \text{ is Col A.}$$

Solution : We write W as a set of linear combinations,

$$W = \left\{ r\begin{bmatrix} 0 \\ 1 \\ 4 \\ 3 \end{bmatrix} + s\begin{bmatrix} 2 \\ 1 \\ 1 \\ -1 \end{bmatrix} + t\begin{bmatrix} 3 \\ -2 \\ 0 \\ -1 \end{bmatrix} : r, s, t \text{ real} \right\}$$

$$= \text{Span} \left\{ \begin{bmatrix} 0 \\ 1 \\ 4 \\ 3 \end{bmatrix}, \begin{bmatrix} 2 \\ 1 \\ 1 \\ -1 \end{bmatrix}, \begin{bmatrix} 3 \\ -2 \\ 0 \\ -1 \end{bmatrix} \right\}$$

Next, use the vectors in spanning set as columns of A.

Let
$$A = \begin{bmatrix} 0 & 2 & 3 \\ 1 & 1 & -2 \\ 4 & 1 & 0 \\ 3 & -1 & -1 \end{bmatrix}$$

Then   W = Col A

We state below the result that we have studied in first semester.

The column space of an $m \times n$ matrix A is all of $\mathbb{R}^m$ if and only if the equation $A\vec{x} = \vec{b}$ has a solution for each $\vec{b}$ in $\mathbb{R}^m$.

Remark : If A is any m × n matrix, then :

1.   The columns space of A is a subspace of $\mathbb{R}^m$.

2.   The null space of A is a subspace of $\mathbb{R}^n$.

3.   If the matrix A is not square, then the Nul A and Col A are completely different, when A is square matrix then zero vector is common to both Nul A and Col A. In this case, in a special case, it is possible that some non-zero vectors belong to both Nul A and Col A.

Example 1.16 : *Let* $A = \begin{bmatrix} -8 & -2 & -9 \\ 6 & 4 & 8 \\ 4 & 0 & 4 \end{bmatrix}$ *and* $\vec{u} = \begin{bmatrix} 2 \\ 1 \\ -2 \end{bmatrix}$. *Determine if* $\vec{u}$ *is in Col A. Is* $\vec{u}$ *in Nul A.*

Solution : For $\vec{u}$ to be in Col A, we need to see that whether $A\vec{x} = \vec{u}$ is consistent. If $A\vec{x} = \vec{u}$ is consistent, means $\vec{u}$ is a linear combination of the columns of A, showing that $\vec{u}$ in Col A, if not consistent, then $\vec{u}$ is not in Col A. To see this, let us consider the augmented matrix $[A\ \vec{u}]$ and row reduce it to row echelon form :

$$[A,\ \vec{u}] = \begin{bmatrix} -8 & -2 & -9 & 2 \\ 6 & 4 & 8 & 1 \\ 4 & 0 & 4 & -2 \end{bmatrix}$$

$$\frac{1}{4}R_3 \sim \begin{bmatrix} -8 & -2 & -9 & 2 \\ 6 & 4 & 8 & 1 \\ 1 & 0 & 1 & -\frac{1}{2} \end{bmatrix}$$

$$R_{13} \sim \begin{bmatrix} 1 & 0 & 1 & -\frac{1}{2} \\ 6 & 4 & 8 & 1 \\ -8 & -2 & -9 & 2 \end{bmatrix}$$

$$\begin{matrix} R_2 + (-4)\,R_1 \\ \sim \\ R_3 + 8R_1 \end{matrix} \begin{bmatrix} 1 & 0 & 1 & -\frac{1}{2} \\ 0 & 4 & 2 & 4 \\ 0 & -2 & -1 & -2 \end{bmatrix}$$

$$R_2 + 2R_3 \sim \begin{bmatrix} 1 & 0 & 1 & -\dfrac{1}{2} \\ 0 & 0 & 0 & 0 \\ 0 & -2 & -1 & -2 \end{bmatrix}$$

$$R_{23} \sim \begin{bmatrix} 1 & 0 & 1 & -\dfrac{1}{2} \\ 0 & -2 & -1 & -2 \\ 0 & 0 & 0 & 0 \end{bmatrix}$$

This shows that the equation $A\vec{x} = \vec{u}$ is consistent, hence $\vec{u}$ is in Col A.

If $\vec{u}$ is in Nul A, then it must satisfy the equation $A\vec{x} = \vec{0}$.

So, consider,

$$A\vec{u} = \begin{bmatrix} -8 & -2 & -9 \\ 6 & 4 & 8 \\ 4 & 0 & 4 \end{bmatrix} \begin{bmatrix} 2 \\ 1 \\ -2 \end{bmatrix}$$

$$= \begin{bmatrix} -16 - 2 + 18 \\ 12 + 4 - 16 \\ 8 - 8 \end{bmatrix} = \begin{bmatrix} 0 \\ 0 \\ 0 \end{bmatrix}$$

Thus, $\vec{u}$ is in Nul A.

Note : The matrix A is square matrix and there is a non-zero vector which is common to both Col A and Nul A.

Important points which differentiate column space and null space for m × n matrix A :

Col A	Nul A
1. Col A is a subspace of $\mathbb{R}^m$.	1. Nul A is subspace of $\mathbb{R}^n$.
2. Col A is explicitly defined, i.e., you are told how to build vectors in Col A.	2. Nul A is implicitly defined, i.e. you are given a condition $A\vec{x} = \vec{0}$, that vectors in Nul A must satisfy.
3. It is easy to find a vector in Col A, which is formed from columns of A.	3. To find vectors in Nul A it is required to reduce [A 0] to row echelon form; this take more time.

... (Contd.)

Col A	Nul A
4. There is obvious relation between Col A and the entries in A, as each column of A is in Col A.	4. There is no obvious relation between Nul A and the entries in A.
5. Any vector $\vec{u}$ in Col A has the property that the equation $A\vec{x} = \vec{u}$ is consistent.	5. Any vector $\vec{u}$ in Nul A has the property that $A\vec{u} = \vec{0}$.
6. Given a vector $\vec{u}$, it takes time to tell if $\vec{u}$ is in Col A. Required row operations on $[A\ \vec{u}]$.	6. Given a vector $\vec{u}$, it is easy to check if $\vec{u}$ is in Nul A. Just compute $A\vec{u}$.
7. Col A $= \mathbb{R}^m$ if and only if the equation $A\vec{x} = \vec{b}$ in is consistent for each $\vec{b}$ in $\mathbb{R}^m$.	7. Nul A $= \{\vec{0}\}$ if and only if the equation $A\vec{x} = \vec{0}$ has only trivial solution.
8. Col A $= \mathbb{R}^m$ if and only if the linear transformation $\vec{x} \to A\vec{x}$ maps $\mathbb{R}^n$ onto $\mathbb{R}^m$.	8. Nul A $= \{0\}$ if and only if the linear transformation $\vec{x} \to A\vec{x}$ is one-to-one.

Kernel and Range of a Linear Transformation :

We generalize the definition of a linear transformation given in Semester - I.

Definitions : A **linear transformation** from a vector space V into a vector space U is a map $T : V \to U$ such that :

(i) $T(\vec{u} + \vec{v}) = T(\vec{u}) + T(\vec{v})$ for all $\vec{u}$ and $\vec{v}$ in V, and

(ii) $T(\alpha\vec{u}) = \alpha T(\vec{u})$ for all $\vec{u}$ in V and all scalars α.

Definition : The **Kernel** (or **null space**) of a linear transformation $T : V \to U$ is the set of all $\vec{u} \in V$ such that $T(\vec{u}) = \vec{0}$ (the zero vector in U). Thus,

$$\text{Kernel } T = \text{Nul A} = \{\vec{u} \in V \mid T(\vec{u}) = \vec{0}\}$$

The **range** of T is the set of all vectors in U of the form $T(\vec{x})$, for $\vec{x}$ in V. That is,

$$R(T) = \text{range of } T = \{\vec{v} \in U \mid T(\vec{x}) = \vec{v}, \text{ for some } \vec{x} \text{ in V}\}$$

Note : If T happens to be a matrix transformation, say, $T(\vec{x}) = A\vec{x}$ for some matrix A, then the Kernel and range of T are just Nul A and Col A.

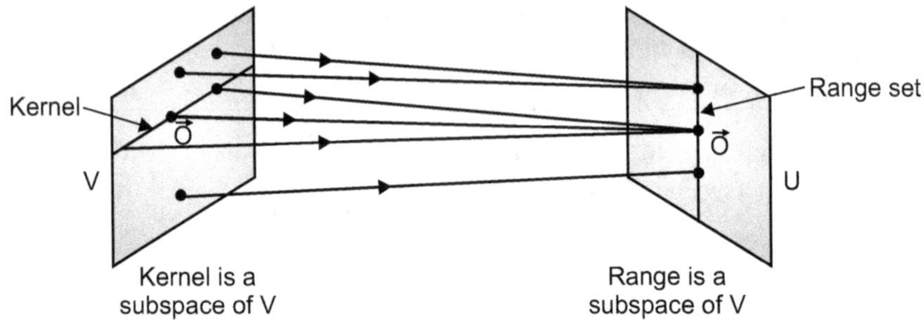

Fig. 1.6 : Subspaces Associated with Linear Transformation

Example 1.17 : *Let V be the set of all real valued functions defined on [a, b] which are differentiable and derivatives are continuous functions on [a, b]. Let W be set of all continuous functions on [a, b].*

Then map D : V → W defined by f(x) ∈ V, $D(f(x)) = \dfrac{d}{dx}(f(x)) = f'(x)$.

Solution: Since, f'(x) is continuous on [a, b], this defines a map D : V → W. The properties of derivatives : Derivative of sum is equal to sum of the derivatives i.e. D(f + g) = D(f) + D(g) for any f, g ∈ V. Also, for any scalar D(αf) = αD(f). This shows that D is a linear transformation.

Note : In above example (16), the Kernel is clearly the set of all constant functions defined on [a, b] and range of D is the set W of all continuous functions on [a, b].

Example 1.18 : *Define $T : P_2 \to \mathbb{R}^2$ by $T(f(x)) = \begin{bmatrix} f(0) \\ f(1) \end{bmatrix}$, f(x) ∈ P_2.*

(a)   Show that T is a linear transformation.

(b)   Find a polynomial p in P_2 that spans the Kernel of T and describe the range of T.

Solution : (a) For f(x), g(x) ∈ P_2, we have

$$T(f(x) + g(x)) = T((f+g)(x)) = \begin{bmatrix} (f+g)(0) \\ (f+g)(1) \end{bmatrix}$$

By definition of addition of function

$$= \begin{bmatrix} (f+g)(0) \\ (f+g)(1) \end{bmatrix}$$

$$= \begin{bmatrix} f(0) + g(0) \\ f(1) + g(1) \end{bmatrix}$$

$$= \begin{bmatrix} f(0) \\ f(1) \end{bmatrix} + \begin{bmatrix} g(0) \\ g(1) \end{bmatrix} \qquad \text{Matrix addition}$$

$$= T(f(x)) + T(g(x))$$

Also, for any scalar α,

$$T(\alpha f(x)) = T((\alpha f)(x))$$

By definition of scalar multiplication in P_2

$$= \begin{bmatrix} (\alpha f)(0) \\ (\alpha f)(1) \end{bmatrix}$$

$$= \begin{bmatrix} \alpha f(0) \\ \alpha f(1) \end{bmatrix}$$

By definition of scalar multiplication

$$= \alpha \begin{bmatrix} f(0) \\ f(1) \end{bmatrix} \quad \text{By scalar multiplication matrices}$$

$$= \alpha T(f(x))$$

Thus, T is a linear transformation.

(b) $p(x)$ is in Kernel of T if, $T(p(x)) = \begin{bmatrix} P(0) \\ P(1) \end{bmatrix} = \begin{bmatrix} 0 \\ 0 \end{bmatrix}$, that is $p(x)$ is in Kernel of T if $P(0) = 0$ and $P(1) = 0$

For $P(x) \in P_2$, let $P(x) = a_0 + a_1 x + a_2 x^2$ and $p(x)$ is in Kernel of T. Then

$$P(0) = 0 \implies a_0 = 0$$

and $P(1) = 0 \implies a_0 + a_1 + a_2 = 0$, that is $a_1 + a_2 = 0$, as $a_0 = 0$.

Thus, any poly $P(x)$ in P_2 is in Kernel of T, then $P(x)$ must be of the form $P(x) = a_1 x + a_2 x^2$, with $a_1 + a_2 = 0$ or $P(x) = a_1 x + (-a_1) x^2$, for an a_1 as scalar.

Thus $P(x)$ spans Kernel of T, where $P(x) = x - x^2 = x(1 - x)$, since all scalar multiples of $P(x)$ must be in Kernel of T.

To find range of T.

For any $\begin{bmatrix} a \\ b \end{bmatrix}$ in $\mathbb{R}^2$, we have $P(x)$ in P_2 such that $P(0) = a$ and $P(1) = b$, that is we have $P(x) = a + (b - a) x^2$, thus for each $\begin{bmatrix} a \\ b \end{bmatrix}$ in $\mathbb{R}^2$ there is a polynomial $P(x) = a + (b - a) x^2$ in $\mathbb{R}^2$ such that $T(P(x)) = \begin{bmatrix} a \\ b \end{bmatrix}$. Therefore, range of T is $\mathbb{R}^2$.

1.4 Linearly Independent Sets; Bases

The idea of linear independence as defined in $\mathbb{R}^n$ in first semester is to be generalized for a vector space. Further, to study the subsets that span a vector space V or a subspace. In this section, we see these concepts and the related results. First we formally define the concept of linear independence in a vector space V.

Definition : An indexed set of vectors $\{\vec{u}_1, \vec{u}_2, ..., \vec{u}_k\}$ in a vector space V is said to be **linear independent** if the vector equation

$$\alpha_1 \vec{u}_1 + \alpha_2 \vec{u}_2 + ... + \alpha_k \vec{u}_k = \vec{0} \qquad ... (1)$$

has only trival solution, $\alpha_1 = 0, ..., \alpha_k = 0$.

The set $\{\vec{u}_1, ..., \vec{u}_k\}$ is said to be **linearly dependent** if (1) has a non-trivial solution, that is, if there exist some weights $\alpha_1, \alpha_2, ..., \alpha_k$, not all zero, such that (1) holds. In this case, (1) is called a **linear dependence relation** among $\vec{u}_1, ..., \vec{u}_k$.

Remark : The simple consequence of the definition that we have seen in $\mathbb{R}^n$, we recall these here :

1. The set containing a single vector $\vec{u}$ is linearly independent if and only if $\vec{u} \neq \vec{0}$.

2. A set of two vectors is linearly dependent if and only if one of the vectors is a scalar multiple of the other.

3. Any set which contains a zero vector is always linearly dependent.

In the following, we prove the theorem which is proved in Semester - I, for a vector space.

Theorem 5 An indexed sets $= \{\vec{u}_1, ..., \vec{u}_k\}$ of two or more vectors, with $\vec{u}_1 \neq \vec{0}$, linearly dependent if and only if some $\vec{u}_i$, $i > 1$ is a linear combination of the preceding vectors $\vec{u}_1, ..., \vec{u}_{i-1}$.

Proof : Suppose the S is linearly dependent. Since, $\vec{u}_1 \neq \vec{0}$, and there exist weights $\alpha_1, ..., \alpha_k$, not all zero, such that

$$\alpha_1 \vec{u}_1 + \alpha_2 \vec{u}_2 + ... + \alpha_k \vec{u}_k = \vec{0}.$$

Let i be the largest subscript for which $\alpha_i \neq 0$, i.e. $\alpha_{i+1} = 0, \ldots, \alpha_k = 0$. If $i = 1$, then $\alpha_1 \vec{u}_1 = \vec{0}$ and $\alpha_1 \neq 0$, so $\vec{u}_1 = \vec{0}$ which is impossible because $\vec{u}_1 \neq \vec{0}$. So $i > 1$, and

$$\alpha_1 \vec{u}_1 + \ldots + \alpha_i \vec{u}_i + 0.\vec{u}_{i+1} + \ldots + 0.\vec{u}_k = \vec{0}$$

This gives $\alpha_i \vec{u}_i = -\alpha_1 \vec{u}_1 \ldots \alpha_{i-1} \vec{u}_{i-1}$

or $\vec{u}_i = \left(-\dfrac{\alpha_1}{\alpha_i}\right) \vec{u}_1 + \ldots + \left(\dfrac{-\alpha_{i-1}}{\alpha_i}\right) \vec{u}_{i-1}.$

Showing that $\vec{u}_i$ is linear combination of the preceding vectors $\vec{u}_1, \ldots, \vec{u}_{i-1}$.

Conversely, suppose one of the vectors say $\vec{u}_i$, $1 < i \leq k$ is linear combination of the preceding vectors $\vec{u}_1, \ldots, \vec{u}_{i-1}$, that is

$$\vec{u}_i = \alpha_1 \vec{u}_1 + \ldots + \alpha_{i-1} \vec{u}_{i-1}, \text{ where } \alpha_1, \ldots, \alpha_{i-1} \text{ scalars.}$$

This can be written as

$$\alpha_1 \vec{u}_1 + \ldots + \alpha_{i-1} \vec{u}_{i-1} + (-1) \vec{u}_i = 0 \qquad \ldots (*)$$

(By adding $(-1) \vec{u}_i$ on both sides)

Clearly (*) is a linear dependence relation with non-zero weight (-1) of $\vec{u}_i$. Therefore, the set s is linearly dependent. ∎

Illustrative Examples

Example 1.19 : *The function 1, $\cos^2 x$, $\sin^2 x$ defined on $\mathbb{R}$, are linearly dependent.*

Solution : Suppose, we have scalars α_1, α_2 and α_3 such that

$$\alpha_1(1) + \alpha_2 \cos^2 x + \alpha_3 \sin^2 x = 0, \ \forall\, x \in \mathbb{R}$$

We know that $\cos^2 x + \sin^2 x = 1$, so taking $\alpha_2 = 1 = \alpha_3$ and $\alpha_1 = -1$, we get

$$(-1)(1) + \cos^2 x + \sin^2 x = 0 \qquad \ldots (*)$$
$$-1 + 1 = 0$$
$$0 = 0$$

Thus, $\alpha_1 = -1$, $\alpha_2 = 1$, $\alpha_3 = 1$, (*) is a linear dependence relation, therefore 1, $\cos^2 x$, $\sin^2 x$ are linearly dependent.

Example 1.20 : *Show that cos x, sin 2x, cos 2x are linearly independent for all $x \in \mathbb{R}$.*

Solution : Suppose we have scalars α_1, α_2 and α_3 such that

$$\alpha_1 \cos x + \alpha_2 \sin 2x + \alpha_3 \cos 2x = 0 \qquad \ldots (1)$$

By taking $x = 0$, in equation (1), we get $\forall\, x \in \mathbb{R}$

$$\alpha_1 + \alpha_3 = 0$$

By taking $x = \dfrac{\pi}{4}$ in equation (1), we get

$$\alpha_1 \frac{1}{\sqrt{2}} + \alpha_2 = 0$$

And by taking $x = \dfrac{\pi}{2}$, in equation (1) we get,

$$\alpha_1 \cdot 0 + \alpha_2 \cdot 0 + \alpha_3\,(-1) = 0$$

That is $\alpha_3 = 0$, using this value, we get $\alpha_1 = 0$, $\alpha_2 = 0$, $\alpha_3 = 0$.

Therefore, the functions are linearly independent.

Example 1.21 : *The polynomial functions $P_1(x) = 1 + x^2$ and $P_2(x) = 1 - x^2$ are linearly independent on $(-\infty, \infty)$.*

Solution : Suppose, we have scalars

$$\alpha P_1(x) + \beta P_2(x) = 0, \ \text{ that is } \ \alpha(1 + x^2) + \beta(1 - x^2) = 0$$

$$\Rightarrow \quad (\alpha + \beta) + (\alpha - \beta)\, x^2 = 0, \ \forall\, x \in \mathbb{R}.$$

This implies that $\alpha + \beta = 0$ and $\alpha - \beta = 0$, giving $\alpha = 0$, $\beta = 0$.

(Alternatively one can see that if $x = 1$

$$\alpha + \beta + \alpha - \beta = 0, \ \text{i.e.,} \ \alpha = 0 \ \text{and} \ x = 0$$

$$\alpha + \beta = 0, \ \text{so} \ \beta = 0 \ \text{etc.})$$

Therefore, $P_1(x)$ and $P_2(x)$ are linearly independent.

Example 1.22 : The polynomial functions defined as $P_1(x) = 1 + x$, $P_2(x) = 1 - x$ and $P_3(x) = 2$, for $x \in (-\infty, \infty)$. It is clearly (obvious) that

$$P_1(x) + P_2(x) - P_3(x) = (1 + t) + (1 - t) + (-2) = 0$$

So, $P_1(x) + P_2(x) - P_3(x)$ is linear dependence relation, hence the functions are linearly dependent.

Definition : Let W be a subspace of a vector space V. An indexed set $\beta = \{\vec{u}_1, \ldots, \vec{u}_k\}$ in V is said to be a **basis** for W if :

(i) β is linearly independent set, and

(ii)   The subspace spanned by β is W, that is W = Span $\{\vec{u}_1, \ldots, \vec{u}_k\}$

Note :

1. If $W = V$, then the definition of a basis for V applies, since any vector space is a subspace of itself.

2. Thus a **basis** of V is a linearly independent set that spans V.

Example 1.23 : (1) The columns of the 3×3 identity matrix $\vec{e_1}$, $\vec{e_2}$, $\vec{e_3}$, where $\vec{e_1} = \begin{bmatrix} 1 \\ 0 \\ 0 \end{bmatrix}$, $\vec{e_2} = \begin{bmatrix} 0 \\ 1 \\ 0 \end{bmatrix}$, $\vec{e_3} = \begin{bmatrix} 0 \\ 0 \\ 1 \end{bmatrix}$ form a basis of $\mathbb{R}^3$.

(2) In general the columns of $n \times n$ matrix; $\vec{e_1}$, $\vec{e_2}$, ..., $\vec{e_n}$, where,

$$\vec{e_1} = \begin{bmatrix} 1 \\ 0 \\ \vdots \\ 0 \end{bmatrix}, \quad \vec{e_2} = \begin{bmatrix} 0 \\ 1 \\ 0 \\ \vdots \\ 0 \end{bmatrix}, \quad \dots \quad \vec{e_n} = \begin{bmatrix} 0 \\ 0 \\ \vdots \\ 1 \end{bmatrix}$$

form a basis of $\mathbb{R}^n$, which is called a **standard basis** for $\mathbb{R}^n$.

(3) If A is an invertible $n \times n$ matrix, say

$A = [\vec{u_1}, \vec{u_2}, ..., \vec{u_n}]$, then the columns of A form a basis for $\mathbb{R}^n$, because they are linearly independent and span $\mathbb{R}^n$.

Example 1.24 : *Let* $\vec{u_1} = \begin{bmatrix} 1 \\ 0 \\ 0 \end{bmatrix}$, $\vec{u_2} = \begin{bmatrix} 1 \\ 1 \\ 0 \end{bmatrix}$, $\vec{u_3} = \begin{bmatrix} 1 \\ 1 \\ 1 \end{bmatrix}$.

Determine if $\{\vec{u_1}, \vec{u_2}, \vec{u_3}\}$ *is a basis of* $\mathbb{R}^3$.

Solution : Since, there are exactly three vectors here in $\mathbb{R}^3$ we can use several methods. We give two methods here :

Method - 1 : The matrix $A = [\vec{u_1} \ \vec{u_2} \ \vec{u_3}]$ is in echelon form and has three pivot positions, hence the matrix is invertible, therefore, $\{\vec{u_1}, \vec{u_2}, \vec{u_3}\}$ is a basis for $\mathbb{R}^3$.

Method - 2 : First, we show that $\{\vec{u_1}, \vec{u_2}, \vec{u_3}\}$ is linearly independent. If suppose $\alpha_1 \vec{u_1} + \alpha_2 \vec{u_2} + \alpha_3 \vec{u_3} = \vec{0}$.

$$\Rightarrow \quad \alpha_1 \begin{bmatrix} 1 \\ 0 \\ 0 \end{bmatrix} + \alpha_2 \begin{bmatrix} 1 \\ 1 \\ 0 \end{bmatrix} + \alpha_3 \begin{bmatrix} 1 \\ 1 \\ 1 \end{bmatrix} = \begin{bmatrix} 0 \\ 0 \\ 0 \end{bmatrix}$$

$$\Rightarrow \quad \begin{bmatrix} \alpha_1 + \alpha_2 + \alpha_3 \\ \alpha_2 + \alpha_3 \\ \alpha_3 \end{bmatrix} = \begin{bmatrix} 0 \\ 0 \\ 0 \end{bmatrix}$$

This gives $\alpha_1 + \alpha_2 + \alpha_3 = 0$, $\alpha_2 + \alpha_3 = 0$ and $\alpha_3 = 0$, clearly, we get $\alpha_1 = 0$, $\alpha_2 = 0$, $\alpha_3 = 0$.

$\therefore$ $\quad \{\vec{u}_1, \vec{u}_2, \vec{u}_3\}$ is linearly independent.

To show that $\{\vec{u}_1, \vec{u}_2, \vec{u}_3\}$ span $\mathbb{R}^3$.

For any $\vec{u} = \begin{bmatrix} a \\ b \\ c \end{bmatrix}$ in $\mathbb{R}^3$, suppose

$$\vec{u} = \alpha_1 \vec{u}_1 + \alpha_2 \vec{u}_2 + \alpha_3 \vec{u}_3$$

$$\Rightarrow \quad \begin{bmatrix} a \\ b \\ c \end{bmatrix} = \begin{bmatrix} \alpha_1 + \alpha_2 + \alpha_3 \\ \alpha_2 + \alpha_3 \\ \alpha_3 \end{bmatrix}$$

$$\Rightarrow \quad \alpha_1 + \alpha_2 + \alpha_3 = a$$

$$\alpha_2 + \alpha_3 = b$$

$$\alpha_3 = c$$

Solving for α_1, α_2 and α_3, we get $\alpha_1 = a - b$, $\alpha_2 = b - c$, $\alpha_3 = c$.

Thus, any $\vec{u}$ in $\mathbb{R}^3$ can be expressed as a linear combination of $\vec{u}_1, \vec{u}_2, \vec{u}_3$. Hence, $\{\vec{u}_1, \vec{u}_2, \vec{u}_3\}$ span $\mathbb{R}^3$.

Thus, $\{\vec{u}_1, \vec{u}_2, \vec{u}_3\}$ is a basis of $\mathbb{R}^3$.

Example 1.25 : *Prove that, the set $S = \{1, x, x^2, \ldots, x^n\}$ is a basis for* P_n.

Solution : Suppose we have scalars $\alpha_0, \alpha_1, \ldots, \alpha_n$ such that

$$\alpha_0(1) + \alpha_1 x + \alpha_2 x^2 + \ldots + \alpha_n x^n = 0 = 0\,(x) \qquad \ldots (1)$$

the zero polynomial.

Equality in equation (1) shows that the polynomial on the left has the same value as the zero polynomial on the right. The fundamental theorem of algebra says that the only polynomial in P_n with more than n zeros is the zero polynomial. Thus, equation (1) holds only of $\alpha_0 = 0, \ldots, \alpha_n = 0$. This proves S is linearly independent. Also any polynomial of degree $\leq$ n is linear combination of the elements in S, so S is a basis for P_n.

$\boxed{\text{Theorem 6}}$ The Spanning Set Theorem

Let $S = \{\vec{u}_1, ..., \vec{u}_k\}$ be a set in a vector space V, and

Let $W = \text{Span} \{\vec{u}_1, ..., \vec{u}_k\}$:

(a) If one of the vectors in S-say $\vec{u}_i$ is a linear combination of the remaining vectors in S, then the set S_1 formed from S by removing $\vec{u}_i$ still spans W.

(b) If $W \neq \{\vec{0}\}$, some subset of S is a basis for H.

Proof : (a) By rearranging the list of vectors in S, if necessary, we may suppose that $\vec{u}_k$ is a linear combination of $\vec{u}_1, ..., \vec{u}_{k-1}$.

So suppose, $\qquad \vec{u}_k = \alpha_1 \vec{u}_1 + ... + \alpha_{k-1} \vec{u}_{k-1}$ $\qquad$ ... (i)

Since, $W = \text{Span} \{\vec{u}_1, \vec{u}_2, ..., \vec{u}_k\}$, any $\vec{x}$ in W can be written as :

$$\vec{x} = \beta \vec{u}_1 + \beta_2 \vec{u}_2 + ... + \beta_{k-1} \vec{u}_{k-1} + \beta_k(\alpha_1 \vec{u}_1 + ... + \alpha_{k-1} \vec{u}_{k-1})$$

Using vector space axioms, we obtain

$$\vec{x} = (\beta_1 + \beta_k \alpha_1) \vec{u}_1 + ... + (\beta_{k-1} + \beta_k \alpha_{k-1}) \vec{u}_{k-1}$$

Showing that $\vec{x}$ is linear combination of the vectors $\vec{u}_1, ..., \vec{u}_{k-1}$. That is,

$$W = \text{Span} \{\vec{u}_1, ..., \vec{u}_{k-1}\}.$$

(b) If the set S is itself linearly independent, S itself is a basis for W. If S is linearly dependent, then one of the vectors in is a linear combination of the remaining vectors. This vector may be deleted from S and still we retain a set S_1 as spanning set for W. If S_1 has two or more vectors we can repeat this process until the spanning set is linearly independent, and hence is a basis for W. If the spanning set is eventually reduced to one vector, that vector will be non-zero, hence linearly independent, as $W \neq \{\vec{0}\}$. ■

Example 1.26 : *Let $S = \{\vec{u}_1, \vec{u}_2, \vec{u}_3, \vec{u}_n\}$ be set in $\mathbb{R}^3$, where,*

$$\vec{u}_1 = \begin{bmatrix} 1 \\ -3 \\ 0 \end{bmatrix}, \quad \vec{u}_2 = \begin{bmatrix} -2 \\ 9 \\ 0 \end{bmatrix}, \quad \vec{u}_3 = \begin{bmatrix} 0 \\ 0 \\ 0 \end{bmatrix}, \quad \vec{u}_4 = \begin{bmatrix} 0 \\ -3 \\ 5 \end{bmatrix}.$$

Is the set S a basis for $\mathbb{R}^3$? If not is there a subset of S which is a basis for $\mathbb{R}^3$. Justify.

Solution : Since, zero vector $\vec{u}_3$ belongs to S, so S cannot be linearly independent. By theorem (5), we delete $\vec{u}_3$ and retain a set $S_1 = \{\vec{u}_1, \vec{u}_2, \vec{u}_4\}$, such that Span S = Span S_1.

Now the vectors in S_1 are linearly independent. Because, if

$$\alpha_1 \begin{bmatrix} 1 \\ -3 \\ 0 \end{bmatrix} + \alpha_2 \begin{bmatrix} -2 \\ 9 \\ 0 \end{bmatrix} + \alpha_3 \begin{bmatrix} 0 \\ -3 \\ 5 \end{bmatrix} = \begin{bmatrix} 0 \\ 0 \\ 0 \end{bmatrix}$$

$$\Rightarrow \begin{bmatrix} \alpha_1 - 2\alpha_2 \\ -3\alpha_1 + 9\alpha_2 + (-3)\,\alpha_3 \\ 5\alpha_3 \end{bmatrix} = \begin{bmatrix} 0 \\ 0 \\ 0 \end{bmatrix}$$

$$\Leftrightarrow \qquad \alpha_1 - 2\alpha_2 = 0$$
$$-3\alpha_1 + 9\alpha_2 - 3\alpha_3 = 0$$
$$5\alpha_3 = 0$$

This gives, $\quad \alpha_1 - 2\alpha_2 = 0$ and

$$-3\alpha_1 + 9\alpha_2 = 0 \qquad\qquad \because \alpha_3 = 0$$

Solving these two equations, we get

$$a_1 = 0, \ \alpha_2 = 0, \ \alpha_3 = 0$$

Using the fact, that $\vec{u}_1, \vec{u}_2, \vec{u}_4$ linearly independent, so the columns of the matrix $A = [\vec{u}_1, \vec{u}_2, \vec{u}_4]$ span whole of $\mathbb{R}^3$.

$\therefore \quad$ S_1 is a basis for $\mathbb{R}^3$.

Bases for Nul A and Col A :

We know how to find vectors that span the null space of a matrix A. In that case, we find a parametric solution of the equation $A\vec{x} = \vec{0}$, and we express the solution as a linear combination of the vectors with weights as free variables and these vectors span and the null space and are linearly independent, so it is a basis for Nul A.

To find the basis for the column space of a matrix A, the following theorem suggest the way of finding basis for Col A. We use the following result.

Elementary row operations on a matrix do not affect the linear dependence relations among the columns of the matrix.

$\boxed{\textbf{Theorem 7}}$ The pivot columns of a matrix A form a basis for Col A.

Proof : We know that Col A is spanned by the columns of A, but the columns of A may not be linearly independent. To find the spanning set of columns of A which is also linearly independent, we reduce the matrix A to its row echelon form - say B. We know that the set of pivot columns of B is linearly independent, for no vector in the set is a linear combination of the vectors that precede it. As A is row equivalent to B, the pivot columns of A are linearly independent as well, because any linear dependence relation among the columns of A corresponds to a linear dependence relation among the columns of B. For this reason, every non-pivot column of A is a linear combination of the pivot columns of A. The non-pivot columns of A may be deleted from the spanning set for Col A, by the spanning set Theorem. This leaves the pivot columns of A as a basis for Col A. ∎

Example 1.27 : *Let A is row equivalent to B, find the bases for Nul A and Column B, where,*

$$A = \begin{bmatrix} 1 & 2 & -5 & 11 & -3 \\ 2 & 4 & -5 & 15 & 2 \\ 1 & 2 & 0 & 4 & 5 \\ 3 & 6 & -5 & 19 & -2 \end{bmatrix}, \quad B = \begin{bmatrix} 1 & 2 & 0 & 4 & 5 \\ 0 & 0 & 1 & -7/5 & -8/5 \\ 0 & 0 & 0 & 0 & 1 \\ 0 & 0 & 0 & 0 & 0 \end{bmatrix}$$

$$\qquad\quad \uparrow \quad \uparrow \quad \uparrow \quad \uparrow \quad \uparrow \qquad\qquad \uparrow \quad \uparrow \quad \uparrow \quad \uparrow \quad \uparrow$$
$$\qquad\quad \vec{a_1} \ \vec{a_2} \ \vec{a_3} \ \vec{a_4} \ \vec{a_5} \qquad\qquad \vec{b_1} \ \vec{b_2} \ \vec{b_3} \ \vec{b_4} \ \vec{b_5}$$

Solution : The matrix B is row-echelon form of A. We observe that each non-pivot column of B is linear combination of the pivot columns.

In fact $\vec{b_2} = 2\vec{b_1}$ and $\vec{b_4} = 4\vec{b_1} + \left(\dfrac{-7}{5}\right)\vec{b_3}$ ... (i)

By spanning Set Theorem, we delete $\vec{b_2}$, $\vec{b_4}$ and $\{\vec{b_1}, \vec{b_3}, \vec{b_4}\}$ will still span Column B. Let

$$S = \{\vec{b_1}, \vec{b_3}, \vec{b_5}\} = \left\{ \begin{bmatrix} 1 \\ 0 \\ 0 \\ 0 \end{bmatrix}, \begin{bmatrix} 0 \\ 1 \\ 0 \\ 0 \end{bmatrix}, \begin{bmatrix} 5 \\ 8/5 \\ 1 \\ 0 \end{bmatrix} \right\}$$

As $\vec{b_1} \neq \vec{0}$ and no vector is S is linear combination of the vectors that precede it, S is linearly independent. Thus, S is a basis for Column B.

Since, A is a row equivalent to B, we have from (i), the corresponding linear dependence relations.

$$\vec{a_2} = 2\vec{a_1} \text{ and } \vec{a_4} = 4\vec{a_1} + \left(\frac{-7}{5}\right)\vec{a_3} \qquad \ldots \text{(ii)}$$

This is by theorem (6).

Thus, we delete the vectors $\vec{a_2}$ and $\vec{a_4}$, when selecting a minimal spanning set for Col A. In fact, $\{\vec{a_1}, \vec{a_3}, \vec{a_5}\}$ is linearly independent and is a minimal spanning set, therefore $\{\vec{a_1}, \vec{a_3}, \vec{a_5}\}$ is a basis for Col A.

Now to find basis for null space of A, we use to express the parametric solution of $A\vec{x} = \vec{0}$.

Thus, writing the corresponding system of linear equations of B, we have

$$x_1 + 2x_2 + 4x_2 + 5x_5 = 0$$

$$x_3 - \frac{7}{5}x_4 + 8x_5 = 0$$

$$x_5 = 0$$

This gives, $\qquad x_1 = -2x_2 - 4x_4$

$$x_3 = \frac{7}{5}x_4$$

with x_1, x_3, x_5 as basic variables and x_2 and x_4 as free variables. The generation is obtained as :

$$\begin{bmatrix} x_1 \\ x_2 \\ x_3 \\ x_4 \\ x_5 \end{bmatrix} = \begin{bmatrix} -2x_2 - 4x_4 \\ 1 \\ \frac{7}{5}x_4 \\ x_4 \\ 0 \end{bmatrix} = x_2 = \begin{bmatrix} -2 \\ 1 \\ 0 \\ 0 \\ 0 \end{bmatrix} + x_4 \begin{bmatrix} -4 \\ 0 \\ \frac{7}{5} \\ 1 \\ 0 \end{bmatrix} = x_2\vec{u} + x_4\vec{v}$$

$$\qquad\qquad\qquad\qquad\qquad\qquad\qquad \uparrow \qquad\qquad \uparrow$$
$$\qquad\qquad\qquad\qquad\qquad\qquad\qquad \vec{u} \qquad\qquad \vec{v}$$

Thus, Nul A is spanned by $\vec{u}$ and $\vec{v}$.

It is easy to see that $\vec{u}$ and $\vec{v}$ are linearly independent, hence $\{\vec{u}, \vec{v}\}$ is basis for Nul A.

Note : IF B is row echelon form of a matrix A, the columns of A which correspond to pivot columns of B form a basis for Col A, not the pivot columns of B. In example (1.26) above all the pivot columns of B have a zero entry in fourth place hence **cannot** span $\mathbb{R}^4$.

Remark : We can view a basis for a vector space V in two ways :

(1) Using spanning set theorem, we delete the vectors from spanning set, this must be stopped as soon as it becomes linearly independent if an additional vector is deleted it will not be linear combination of the remaining vectors, hence the smaller set will no longer span V. **Thus, a basis is a spanning set that is as small as possible.**

(2) **Basis is also a linearly independent set in V that is as large as possible.** If S is a basis for V, and if S is enlarged by one vector - say $\vec{v}$ from V, then the new set cannot be linearly independent, as S spans V.

Example 1.28 : Let us consider the following three sets in $\mathbb{R}^3$.

$$\left\{ \begin{bmatrix} 1 \\ 0 \\ 0 \end{bmatrix}, \begin{bmatrix} -2 \\ 1 \\ 0 \end{bmatrix} \right\} \quad \left\{ \begin{bmatrix} 1 \\ 0 \\ 0 \end{bmatrix}, \begin{bmatrix} -2 \\ 1 \\ 0 \end{bmatrix}, \begin{bmatrix} 0 \\ 5 \\ 1 \end{bmatrix} \right\} \quad \left\{ \begin{bmatrix} 1 \\ 0 \\ 0 \end{bmatrix}, \begin{bmatrix} -2 \\ 1 \\ 0 \end{bmatrix}, \begin{bmatrix} 0 \\ 5 \\ 1 \end{bmatrix}, \begin{bmatrix} 3 \\ 2 \\ 1 \end{bmatrix} \right\}$$

$$\uparrow \qquad\qquad \uparrow \qquad\qquad\qquad \uparrow$$

Linearly independent   a basis for   Spans $\mathbb{R}^3$ but is not

but cannot span $\mathbb{R}^3$ $\mathbb{R}^3$ linearly independent

We see that the first linearly independent set can be enlarged to a basis - second set.

Further enlargement second to third destroys the linearly independence.

Also, we see that the spanning set can be shrunk to a basis (second set from fourth set) to a basis.

1.5 Co-ordinate Systems

In $\mathbb{R}^2$, $\mathbb{R}^3$ or $\mathbb{R}^n$ we have a co-ordinate system and we use usually standard bases of these spaces. In the same way the basis for a vector space imposes a co-ordinate system on V.

Theorem 8 **The Unique Representation Theorem :**

Let $\beta = \{\vec{u}_1, \vec{u}_2, ..., \vec{u}_n\}$ be a basis for a vector space V. Then to each $\vec{x}$ in V, there exist unique scalars $\alpha_1, \alpha_2, ..., \alpha_n$ such that

$$\vec{x} = \alpha_1 \vec{u}_1 + \alpha_2 \vec{u}_2 + \alpha_3 \vec{u}_3 + ... + \alpha_n \vec{u}_n$$

Proof : Since, β is a basis for V, β spans V, so there exist scalars $\alpha_1, \alpha_2, ..., \alpha_n$, such that

$$\vec{x} = \alpha_1 \vec{u}_1 + \alpha_2 \vec{u}_2 + ... + \alpha_n \vec{u}_n \qquad\qquad ... (1)$$

Suppose there exist scalars $\beta_1, \beta_2, \ldots, \beta_n$ such that

$$\vec{x} = \beta_1\vec{u}_1 + \beta_2\vec{u}_2 + \ldots + \beta_n\vec{u}_n \qquad \ldots (2)$$

Subtracting equation (2) from equation (1), we get

$$\vec{0} = \vec{x} - \vec{x} = (\alpha_1 - \beta_1)\,\vec{u}_1 + \ldots + (\alpha_n - \beta_n)\,\vec{u}_n \qquad \ldots (3)$$

Here we are using vector space axiom to get equation (3).

Now β is a basis for V, so is linearly independent, hence the weight in equation (3) must all be zero, that is

$$\alpha_1 - \beta_1 = 0, \quad \alpha_2 - \beta_2 = 0, \ldots, \alpha_n - \beta_n = 0$$

That is $\alpha_1 = \beta_1, \quad \alpha_2 - \beta_2, \ldots, \alpha_n = \beta_n$ ∎

Definition : Let $\beta = \{\vec{u}_1, \vec{u}_2, \ldots, \vec{u}_n\}$ be a basis for V and x be a vector in V.

Then the unique representation

$$\vec{x} = \alpha_1\vec{u}_1 + \alpha_2\vec{u}_2 + \ldots + \alpha_n\vec{u}_n$$

with scalars $\alpha_1, \alpha_2, \ldots, \alpha_n$ holds. The weights (scalars) $\alpha_1, \alpha_2, \ldots, \alpha_n$ uniquely determined by a basis β for $\vec{x}$ are called **the co-ordinates of $\vec{x}$ relative to the basis β (or the β-co-ordinates of $\vec{x}$).**

If $\alpha_1, \alpha_2, \ldots, \alpha_n$ are β-co-ordinates of $\vec{x}$, then the vector in $\mathbb{R}^n$.

$$[\vec{x}]_\beta = \begin{bmatrix} \alpha_1 \\ \alpha_2 \\ \vdots \\ \alpha_n \end{bmatrix}$$

is called **the co-ordinate vector of $\vec{x}$ relative to β, or the β-co-ordinate vector of $\vec{x}$.**

Note : From above definition, we see that there is a map defined from V to $\mathbb{R}^n$ which maps each $\vec{x}$ in V to the unique vector $[\vec{x}]_\beta$ in $\mathbb{R}^n$, $\vec{x} \to [\vec{x}]_\beta$. This map is called the **co-ordinate mapping** determined by β.

Example 1.29 : *Find the vector $\vec{x}$ in (a) and (b) determined by the given co-ordinate vector $[\vec{x}]_\beta$ and given basis :*

(a) $\beta = \left\{ \begin{bmatrix} 4 \\ 5 \end{bmatrix}, \begin{bmatrix} 6 \\ 7 \end{bmatrix} \right\}, \quad [\vec{x}]_\beta = \begin{bmatrix} 8 \\ -5 \end{bmatrix}$

(b) $\beta = \left\{ \begin{bmatrix} 1 \\ -4 \\ 3 \end{bmatrix}, \begin{bmatrix} 5 \\ 2 \\ -2 \end{bmatrix}, \begin{bmatrix} 4 \\ -7 \\ 0 \end{bmatrix} \right\}$, $[\vec{x}]_\beta = \begin{bmatrix} 3 \\ 0 \\ -1 \end{bmatrix}$

(c)   Find co-ordinate vector of $\vec{x} = \begin{bmatrix} 2 \\ 5 \end{bmatrix}$ with respective to the basis

$\beta = \left\{ \begin{bmatrix} 4 \\ 5 \end{bmatrix}, \begin{bmatrix} 6 \\ 7 \end{bmatrix} \right\}$.

(d)   Given $\vec{x} = \begin{bmatrix} -1 \\ -5 \\ 9 \end{bmatrix}$, find $[\vec{x}]_\beta$, where β is basis as given in (b).

Solution : (a) By the Unique Representation Theorem,

$$\vec{x} = 8 \begin{bmatrix} 4 \\ 5 \end{bmatrix} + (-5) \begin{bmatrix} 6 \\ 7 \end{bmatrix} = \begin{bmatrix} 32 \\ 40 \end{bmatrix} + \begin{bmatrix} -30 \\ -35 \end{bmatrix} = \begin{bmatrix} 2 \\ 5 \end{bmatrix}$$

(b) $\vec{x} = 3 \begin{bmatrix} 1 \\ -4 \\ 3 \end{bmatrix} + 0 \begin{bmatrix} 5 \\ 2 \\ -2 \end{bmatrix} + (-1) \begin{bmatrix} 4 \\ -7 \\ 0 \end{bmatrix} = \begin{bmatrix} 3 \\ -12 \\ 9 \end{bmatrix} + \begin{bmatrix} 0 \\ 0 \\ 0 \end{bmatrix} + \begin{bmatrix} -4 \\ 7 \\ 0 \end{bmatrix} = \begin{bmatrix} -1 \\ -5 \\ 9 \end{bmatrix}$.

(c)   Suppose, $\vec{x} = \alpha_1 \begin{bmatrix} 4 \\ 5 \end{bmatrix} + \alpha_2 \begin{bmatrix} 6 \\ 7 \end{bmatrix}$

     That is $\begin{bmatrix} 2 \\ 5 \end{bmatrix} = \alpha_1 \begin{bmatrix} 4 \\ 5 \end{bmatrix} + \alpha_2 \begin{bmatrix} 6 \\ 7 \end{bmatrix}$

$\Rightarrow$ $\begin{bmatrix} 2 \\ 5 \end{bmatrix} = \begin{bmatrix} 4\alpha_1 + 6\alpha_2 \\ 5\alpha_1 + 7\alpha_2 \end{bmatrix}$

$\Leftrightarrow$ $4\alpha_1 + 6\alpha_2 = 2$ ... (1)

         $5\alpha_1 + 7\alpha_2 = 5$ ... (2)

Multiplying first equation by 5 and second by 4 and subtract, we get

     $2\alpha_2 = -10$ i.e. $\alpha_2 = -5$

and using $\alpha_2 = -5$ in first equation gives $\alpha_1 = 8$. Thus, $[\vec{x}]_\beta = \begin{bmatrix} 8 \\ -5 \end{bmatrix}$.

(d)   Suppose $\vec{x} = \begin{bmatrix} -1 \\ -5 \\ 9 \end{bmatrix} = \alpha_1 \begin{bmatrix} 1 \\ -4 \\ 3 \end{bmatrix} + \alpha_2 \begin{bmatrix} 5 \\ 2 \\ -2 \end{bmatrix} + \alpha_3 \begin{bmatrix} 4 \\ -7 \\ 0 \end{bmatrix}$

$\Rightarrow$ $\begin{bmatrix} -1 \\ -5 \\ 9 \end{bmatrix} = \begin{bmatrix} \alpha_1 + 5\alpha_2 + 4\alpha_3 \\ -4\alpha_1 + 2\alpha_2 - 7\alpha_3 \\ 3\alpha_1 - 2\alpha_2 \end{bmatrix} = \begin{bmatrix} 1 & 5 & 4 \\ -1 & 2 & -7 \\ 3 & -2 & 0 \end{bmatrix} \begin{bmatrix} \alpha_1 \\ \alpha_2 \\ \alpha_3 \end{bmatrix}$

$\Leftrightarrow$ $\alpha_1 + 5\alpha_2 + 4\alpha_3 = -1$

         $-4\alpha_1 + 2\alpha_2 - 7\alpha_3 = -5$

             $3\alpha_1 - 2\alpha_2 = 9$

Solving this system of linear equations, by any one method, by row reduction or eliminated method. We obtain

$$\alpha_1 = 3, \ \alpha_2 = 0, \ \alpha_3 = -1$$

Therefore, $\qquad [\vec{x}]_\beta = \begin{bmatrix} +3 \\ 0 \\ -1 \end{bmatrix}$

Note :

1. In case of $\mathbb{R}^2$, $\mathbb{R}^3$ or $\mathbb{R}^n$, the entries in the vector $\vec{x}$ are the co-ordinates of $\vec{x}$ relative to the standard basis.

    For $\vec{x} = \begin{bmatrix} 3 \\ 5 \end{bmatrix}$ in $\mathbb{R}^2$, $[\vec{x}]_\beta = \begin{bmatrix} 3 \\ 5 \end{bmatrix}$, where, $\beta = \left\{ \begin{bmatrix} 1 \\ 0 \end{bmatrix}, \begin{bmatrix} 0 \\ 1 \end{bmatrix} \right\}$

    For $\vec{x} = \begin{bmatrix} 1 \\ 5 \\ -7 \end{bmatrix}$ in $\mathbb{R}^3$, $[\vec{x}]_\beta = \begin{bmatrix} 1 \\ 5 \\ -7 \end{bmatrix}$, where, $\beta = \left\{ \begin{bmatrix} 1 \\ 0 \\ 0 \end{bmatrix}, \begin{bmatrix} 0 \\ 1 \\ 0 \end{bmatrix}, \begin{bmatrix} 0 \\ 0 \\ 1 \end{bmatrix} \right\}$

    Similar facts can be seen for $\mathbb{R}^n$.

2. In the example 1.29 (d), we see that the β-co-ordinates of the vector $\vec{x}$ are changed into the standard co-ordinates by the matrix formed by the basis vector.

    In general, if $\beta = \{\vec{u}_1, \vec{u}_2, \dots, \vec{u}_n\}$ is a basis for $\mathbb{R}^n$, let the matrix whose columns are basis elements be denoted by P_β, so $P_\beta = [\vec{u}_1, \vec{u}_2, \dots, \vec{u}_n]$. Then the vector equation

    $$\vec{x} = \alpha_1 \vec{u}_1 + \alpha_2 \vec{u}2 + \dots + \alpha_n \vec{u}_n \text{ is equivalent to}$$

    $$\vec{x} = P_\beta [\vec{x}]_\beta \qquad \dots (*)$$

    We call P_β **the change of co-ordinates matrix** from β to the standard basis for $\mathbb{R}^n$. P_β acts as a map from $\mathbb{R}^n$ into $\mathbb{R}^n$, which maps $[\vec{x}]_\beta$ to $\vec{x}$, is a co-ordinate mapping.

    Since, the columns of P_β are the basis elements of $\mathbb{R}^n$, P_β is invertible. Left multiplication by P_β^{-1} in (*) converts $\vec{x}$ into its β-co-ordinate vector :

    $$P_\beta^{-1} \vec{x} = [\vec{x}]_\beta$$

This correspondence is $\vec{x} \to [\vec{x}]_\beta$, produced by P_β^{-1}, is the co-ordinate mapping. Since, P_β^{-1} is invertible, the co-ordinate mapping is a one-to-one linear transformation from $\mathbb{R}^n$ onto $\mathbb{R}^n$. This property of the co-ordinate mapping is true in a general vector space that has a basis.

The Co-ordinate Mapping :

Let V be a vector space and let $\beta = \{\vec{u}_1, \vec{u}_2, ..., \vec{u}_n\}$ be a basis for V. This basis introduces a co-ordinate mapping V. For $\vec{x}$ in V, the co-ordinate mapping $\vec{x} \to [\vec{x}]_\beta$ connects the unfamiliar vector space V to the familiar space $\mathbb{R}^n$. The points V can be now be identified by their "names". See the following Fig. 1.7.

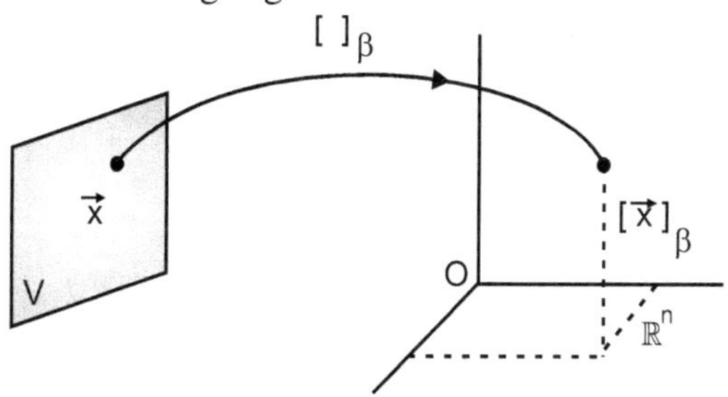

Fig. 1.7

The co-ordinate mapping from V to $\mathbb{R}^n$.

Theorem 9 Let $\beta = \{\vec{u}_1, \vec{u}_2, ..., \vec{u}_n\}$ be a basis for a vector space V. Then the co-ordinate mapping $\vec{x} \to [\vec{x}]_\beta$ is one-to-one linear transformation from V onto $\mathbb{R}^n$.

Proof : Let $\vec{u}$ and $\vec{v}$ be any two vectors in V, then we have (β is a basis)

$$\vec{u} = \alpha_1 \vec{u}_1 + \alpha_2 \vec{u}_2 + ... + \alpha_n \vec{u}_n$$

and $\vec{v} = \beta_1 \vec{u}_1 + \beta_2 \vec{u}_2 + ... + \beta_n \vec{u}_n$

where, α_i's, β_i's $i = 1$ to n are scalars, i.e. real numbers.

Then using vector space axioms, we get

$$\vec{u} + \vec{v} = (\alpha_1 + \beta_1)\, \vec{u}_1 + (\alpha_2 + \beta_2)\, \vec{u}_2 + ... + (\alpha_n + \beta_n)\, \vec{u}_n.$$

It follows that

$$[\vec{u} + \vec{v}]_\beta = \begin{bmatrix} \alpha_1 + \beta_1 \\ \alpha_2 + \beta_2 \\ \vdots \\ \alpha_n + \beta_n \end{bmatrix} = \begin{bmatrix} \alpha_1 \\ \alpha_2 \\ \vdots \\ \alpha_n \end{bmatrix} + \begin{bmatrix} \beta_1 \\ \beta_2 \\ \vdots \\ \beta_n \end{bmatrix} = [\vec{u}]_\beta + [\vec{v}]_\beta$$

This shows that the co-ordinate map preserves addition. Next, if c is any scalar, then

$$\vec{cu} = c(\alpha_1 \vec{u}_1 + \alpha_2 \vec{u}_2 + \ldots + \alpha_n \vec{u}_n)$$

$$= (c\alpha_1)\,\vec{u}_1 + (c\alpha_2)\,\vec{u}_2 + \ldots + (c\alpha_n)\,\vec{u}_n$$

So, we have,
$$[\vec{cu}]_\beta = \begin{bmatrix} c\alpha_1 \\ c\alpha_2 \\ \vdots \\ c\alpha_n \end{bmatrix} = c \begin{bmatrix} \alpha_1 \\ \alpha_2 \\ \vdots \\ \alpha_n \end{bmatrix} = c[\vec{u}]_\beta$$

Showing that the co-ordinate map also preserves the scalar multiplication and hence is a linear transformation.

Now to show that the co-ordinate mapping is one-to-one, suppose $\vec{u}$ and $\vec{v}$ in V have the same images under this map, i.e. $[\vec{u}]_\beta = [\vec{v}]_\beta$

This shows that,

$$\begin{bmatrix} \alpha_1 \\ \alpha_2 \\ \vdots \\ \alpha_n \end{bmatrix} = \begin{bmatrix} \beta_1 \\ \beta_2 \\ \vdots \\ \beta_n \end{bmatrix} \qquad \because [\vec{u}]_\beta = \begin{bmatrix} \alpha_1 \\ \alpha_2 \\ \vdots \\ \alpha_n \end{bmatrix} < [\vec{v}]_\beta = \begin{bmatrix} \beta_1 \\ \beta_2 \\ \vdots \\ \beta_n \end{bmatrix}$$

But in $\mathbb{R}^n$ this holds if and only if $\alpha_i = \beta_i$ for each $i = 1$ to n.

This means $\vec{u} = \vec{v}$.

Therefore, the mapping is one-to-one.

Now to show that the mapping is onto, let $\vec{y} \in \mathbb{R}^n$ be any vector,

where, $\vec{y} = \begin{bmatrix} y_1 \\ y_2 \\ \vdots \\ y_n \end{bmatrix}$, $y_i \in \mathbb{R}$, $i = 1$ to n.

Then we have vector

$$\vec{u} = y_1 \vec{u}_1 + y_2 \vec{u}_2 + \ldots + y_n \vec{u}_n \text{ in V such that}$$

$$[\vec{u}]_\beta = \begin{bmatrix} y_1 \\ y_2 \\ \vdots \\ y_n \end{bmatrix} = \vec{y}. \qquad \blacksquare$$

Remark : If $\vec{v}_1, \vec{v}_2, \ldots, \vec{v}_k$ are vector in V and if $\alpha_1, \alpha_2, \ldots, \alpha_k$ are scalars, then we have a linear combination

$$\alpha_1 \vec{u}_1 + \alpha_2 \vec{u}_2 + \ldots + \alpha_n \vec{u}_n.$$

What is co-ordinate vector of the linear combination relative to the basis β ?

The linearity of the co-ordinate mapping extends to linear combination that is

$$[\alpha_1 \vec{u}_1 + \alpha_2 \vec{u}_2 + \ldots + \alpha_k \vec{u}_k]_\beta = \alpha_1 [\vec{u}_1]_\beta + \alpha_2 [\vec{u}_2]_\beta + \ldots + \alpha_k [\vec{u}_k]_\beta$$

That is β co-ordinate vector of a linear combination of $\vec{u}_1, \ldots, \vec{u}_k$ is the same linear combination of their co-ordinate vectors.

Definition : The one-to-one linear transformation of a vector space V onto a vector space W is called an **isomorphism** from V onto W.

Note : The above co-ordinate mapping is an isomorphism from V onto $\mathbb{R}^n$.

Example 1.30 : The standard basis for P_n is $\beta = \{1, x, x^2, \ldots, x^n\}$. any polynomial in P_n is written as

$$p(x) = a_0 + a_1 x + \ldots + a_n x^n, \ a_0, a_1, \ldots, a_n \in \mathbb{R}.$$

Therefore, $\qquad [p(x)]_\beta = \begin{bmatrix} a_0 \\ a_1 \\ \vdots \\ a_n \end{bmatrix}.$

This shows that the co-ordinate mapping from P_n onto $\mathbb{R}^{n+1}$ is an isomprohism.

Example 1.31 : *Use co-ordinate vectors to verify that the polynomials $1 + x^2$, $x + x^2$, $1 + 2x + x^2$ are linearly independent, hence is a basis for P_2. Find the co-ordinate vector of $p(x) = 1 + 4x + 7x^2$.*

Solution : The co-ordinate vectors of the given polynomials are $(1, 0, 1), (0, 1, 1), (1, 2, 1)$.

Writing these vectors as columns of a matrix A, we can determine their independence by row reducing the augmented matrix for $A\vec{x} = \vec{0}$.

$$\begin{bmatrix} 1 & 0 & 1 & 0 \\ 0 & 1 & 2 & 0 \\ 1 & 1 & 1 & 0 \end{bmatrix} \xrightarrow{R_3 - R_1} \begin{bmatrix} 1 & 0 & 1 & 0 \\ 0 & 1 & 2 & 0 \\ 0 & 1 & 0 & 0 \end{bmatrix}$$

$$\xrightarrow{R_3 - R_2} \begin{bmatrix} 1 & 0 & 1 & 0 \\ 0 & 1 & 2 & 0 \\ 0 & 0 & -2 & 0 \end{bmatrix} \xrightarrow{\frac{1}{2}R_3} \begin{bmatrix} 1 & 0 & 1 & 0 \\ 0 & 1 & 2 & 0 \\ 0 & 0 & 1 & 0 \end{bmatrix}$$

Since, there are three pivot columns the co-ordinate vectors are linearly independent, hence the given polynomials are linearly independent, hence form a basis for P_2.

Now to find the co-ordinate vector for $p(x) = 1 + 4x + 7x^2$, we need to find scalars $\alpha_1, \alpha_2, \alpha_3$ s.t.

$$4x + 7x^2 = \alpha_1(1 + x^2) + \alpha_2(x + x^2) + \alpha_3(1 + 2x + x^2)$$

So that $\alpha_1, \alpha_2, \alpha_3$ are co-ordinates of $p(x)$. This equation reduces to
$1 + 4x + 7x^2 = (\alpha_1 + \alpha_3) + (\alpha_2 + 2\alpha_3) \, x + (\alpha_1 + \alpha_2 + \alpha_3) \, x^2$.

This holds if and only if

$$\alpha_1 + \alpha_3 = 1$$

$$\alpha_2 + 2\alpha_3 = 4$$

$$\alpha_1 + \alpha_2 + \alpha_3 = 7$$

Solving this system by row-reduction :

$$\begin{bmatrix} 1 & 0 & 1 & 1 \\ 0 & 1 & 2 & 4 \\ 1 & 1 & 1 & 7 \end{bmatrix} \xrightarrow{R_3 - R_1} \begin{bmatrix} 1 & 0 & 1 & 1 \\ 0 & 1 & 2 & 4 \\ 0 & 1 & 0 & 6 \end{bmatrix}$$

$$\xrightarrow{R_3 - R_2} \begin{bmatrix} 1 & 0 & 1 & 1 \\ 0 & 1 & 2 & 4 \\ 0 & 0 & -2 & 2 \end{bmatrix} \xrightarrow{\left(\frac{-1}{2}\right) R_3} \begin{bmatrix} 1 & 0 & -1 & 1 \\ 0 & 1 & 2 & 4 \\ 0 & 0 & 1 & -1 \end{bmatrix}$$

Writing the corresponding system, we get

$$\alpha_1 + \alpha_3 = 1$$

$$\alpha_2 + 2\alpha_3 = 4$$

$$\alpha_3 = -1$$

This gives $\alpha_1 = 2, \quad \alpha_2 = 6, \quad \alpha_3 = -1$

$$\therefore \qquad [p(x)]_\beta = \begin{bmatrix} 2 \\ 6 \\ -1 \end{bmatrix}$$

1.6 The Dimension of a Vector Space

We have seen in preceding section that a vector space V with a basis β containing n vectors is isomporphic to $\mathbb{R}^n$. We will show in this section that any two bases for V have the **same number** of vectors.

Theorem 10 If a vector space V has a basis $\beta = \{\vec{u}_1, \vec{u}_2, ..., \vec{u}_n\}$, then any set in V with more that n vectors must be linearly dependent.

Proof : Let $\{\vec{v}_1, \vec{v}_2, ..., \vec{v}_k\}$ be a set in V with $k > n$. The co-ordinate vectors $[\vec{v}_1]_\beta, [\vec{v}_2]_\beta, ..., [\vec{v}_k]_\beta$ are in $\mathbb{R}^n$ and $k > n$.

Therefore $[\vec{v}_1]_\beta$, $[\vec{v}_2]_\beta$, ..., $[\vec{v}_k]_\beta$ are linearly dependent in $\mathbb{R}^n$ because there are more vectors than the entries in each vector. So there exist scalars α_1, α_2, ..., α_k not all zero, such that

$$\alpha_1[\vec{v}_1]_\beta + \alpha_2[\vec{v}_2]_\beta + ..., + \alpha_k[\vec{v}_k]_\beta = \begin{bmatrix} 0 \\ 0 \\ \vdots \\ 0 \end{bmatrix}.$$

As the co-ordinate mapping is linear transformation, the above equation can be written as

$$[\alpha_1\vec{v}_1 + \alpha_2\vec{v}_2 + ... + \alpha_k\vec{v}_k]_\beta = \begin{bmatrix} 0 \\ 0 \\ \vdots \\ 0 \end{bmatrix}, \text{ the zero vector in } \mathbb{R}^n.$$

The zero vector on the right contains n weights needed to build the vector $\alpha_1\vec{v}_1 + \alpha_2\vec{v}_2 + ... + \alpha_n\vec{v}_n$ from the basis vectors in β. That is,

$$\alpha_1\vec{v}_1 + \alpha_2\vec{v}_2 + ... + \alpha_k\vec{v}_k = 0.\vec{u}_1 + 0.\vec{u}_2 + ... + 0.\vec{u}_n = \vec{0}$$

Since, the $\alpha_{i's}$ are not all zero, the set $\{\vec{v}_1, \vec{v}_2, ..., \vec{v}_k\}$ is linearly dependent.

Note : The above theorem implies that if a vector space V has a basis with n-vectors, then each linearly independent set in V has no more than n vectors.

Remark : In a vector space V, an infinite set is said to be linearly dependent if some finite subset is linearly dependent, otherwise the set is linearly independent. If S is an infinite set in V, take any subset $\{\vec{v}_1, \vec{v}_2, ..., \vec{v}_k\}$ of S_1 with $k > n$, if this subset is linearly dependent by using theorem 9 then S is also linearly dependent. ∎

Theorem 11 Let V be a vector space with a basis of n vectors, then every basis of V must consist of exactly n vectors.

Proof : Let $\beta_1 = \{\vec{u}_1, \vec{u}_2, ..., \vec{u}_n\}$ be a basis of V and suppose $\beta_2 = \{\vec{v}_1, \vec{v}_2, ..., \vec{v}_m\}$ be any other basis with m-vectors. Since, β_1 is basis and β_2 is linearly independent, β_2 has no more than n vectors. i.e. $m \leq n$. Also, since β_2 is basis for V and β_1 is linearly independent, β_1 can not have more than m vectors, i.e. $n \leq m$.

Thus, we get $m = n$.

Note : If a non-zero vector space V is spanned by a finite set S, then a subset of S is a basis for V, by the spanning set theorem. ∎

Definition : If a vector space V is spanned by a finite set, then V is said to be **finite-dimensional**, and the **dimension** of V, denoted by dim V, is the number of vectors in a basis for V. The dimension of the zero vector space is $\{\vec{0}\}$ is defined to be zero. If V is not spanned by a finite set, then V is called **infinite dimensional.**

Example : We know that the standard basis for $\mathbb{R}^n$ contains n vectors, so dim $\mathbb{R}^n$ = n. The standard basis for the vector space of polynomials P_n is $\{1, x, x^2, ..., x^n\}$, hence dim P_n = n + 1. In particular dim $\mathbb{R}^2$ = 2, dim $\mathbb{R}^3$ = 3, dim P_2 = 3, dim P_3 = 4. The space P of all polynomials is infinite dimensional.

Subspaces of a Finite Dimensional Space.

$\boxed{\textbf{Theorem 12}}$ Let W be a subspace of finite dimensional vector space V. Then any linearly independent set in W can be expanded, if necessary to a basis for W. Also, W is finite dimensional and dim W $\leq$ dim V.

Proof : If W is trivial subspace of V, that is W = $\{\vec{0}\}$, the certainly dim W = 0 $\leq$ dim V. Suppose W $\neq \{\vec{0}\}$. Let S = $\{\vec{u}_1, \vec{u}_2, ..., \vec{u}_k\}$ be any linearly independent set in W. If S spans W, then S is a basis for W. If S does not span W, there is some vector $\vec{u}_{k+1}$ in W which is not linear combination of vectors in S, that is $\vec{u}_{k+1}$ is not span S. But then the set $S_1 = \{\vec{u}_1, \vec{u}_2, ..., \vec{u}_k, \vec{u}_{k+1}\}$ will be linearly independent, because no vector in the set can be linear combination of the vectors that precede it.

If the new set S_1 does not span W, we can continue this process of expanding S to a larger linearly independent set in W. But the number of vectors in linearly independent expansion of S can never exceed the dimension of V, by theorem 9. So eventually the expansion of S will span W and hence will be a basis for W, and dim W $\leq$ dim V. $\blacksquare$

$\boxed{\textbf{Theorem 13}}$ **(The Basis Theorem) :** Let V be a vector space of dimension n with n $\geq$ 1. Then any linearly independent set with exactly n elements in V is automatically a basis for V. Any set of exactly n elements that span V is automatically a basis for V.

Proof : Suppose S is a set with n linearly independent vectors. Then S can be extended to a basis for V by theorem 11. But that basis must contain exactly n elements, since dim V = n. So S must already be a basis for V.

Now suppose, S is a set of n vectors in V and span S = V. Since, V is not trivial vector space, the spanning set theorem implies that a subset S' of S is a basis for V. Since dim V = n, S' must contain n vectors, hence S' = S. ■

Corollary : Let A be an m × n matrix. Then dim (Nul A) is the number of free variables in the solution of the equation $A\vec{x} = \vec{0}$. The dim (Col A) is the number of pivot columns of A.

Proof : Suppose the equation $A\vec{x} = \vec{0}$ has k free variables, we know that the standard method of finding a spanning set for Nul A will produce exactly k linearly independent vectors - say $\vec{u}_1, \vec{u}_2, ..., \vec{u}_k$ - one for each free variables. So $\{\vec{u}_1, \vec{u}_2, ..., \vec{u}_k\}$ is a basis for Nul A, and the number of free variables determines the size of the basis.

We know that the set of columns of A which are pivot columns is linearly independent and spans the Col A.

Therefore, dim (Col A) = the number of pivot columns of A.

Example 1.32 : *Find the basis and state the dimension of the subspace*

$$W = \left\{ \begin{bmatrix} a + b \\ 2a \\ 3a - b \\ -b \end{bmatrix} : a, b \in \mathbb{R} \right\}$$

Solution : We have, $W = \left\{ \begin{bmatrix} a + b \\ 2a \\ 3a - b \\ -b \end{bmatrix} : a, b \in \mathbb{R} \right\}$

$$= \left\{ a \begin{bmatrix} 1 \\ 2 \\ 3 \\ 0 \end{bmatrix} + b \begin{bmatrix} 1 \\ 0 \\ -1 \\ -1 \end{bmatrix} : a, b \in \mathbb{R} \right\}$$

Every vector in W is linear combination of the vector

$$\vec{u}_1 = \begin{bmatrix} 1 \\ 2 \\ 3 \\ 0 \end{bmatrix} \text{ and } \vec{u}_2 \begin{bmatrix} 1 \\ 0 \\ -1 \\ -1 \end{bmatrix}$$

We see that $\vec{u}_1$ and $\vec{u}_2$ are scalar multiples of each other, hence $\{\vec{u}_1, \vec{u}_2\}$ is a basis and dim W = 2.

Example 1.33 : *Find the basis and dimension of the subspace.*

$$W = \left\{ \begin{bmatrix} 3a + 6b - c \\ 6a - 2b - 2c \\ -9a + 5b + 3c \\ -3a + b + c \end{bmatrix} : a, b, c \in \mathbb{R} \right\}$$

Solution : It is easy to see that each vector in W is a linear combination of the vectors :

$$\vec{u}_1 = \begin{bmatrix} 3 \\ 6 \\ -9 \\ -3 \end{bmatrix}, \quad \vec{u}_2 = \begin{bmatrix} 6 \\ -2 \\ 5 \\ 1 \end{bmatrix} \text{ and } \vec{u}_3 = \begin{bmatrix} -1 \\ -2 \\ 3 \\ 1 \end{bmatrix}$$

That is, W is spanned by $\vec{u}_1$, $\vec{u}_2$ and $\vec{u}_3$.

More observation shows that $\vec{u}_3 = \left(-\dfrac{1}{2}\right) \vec{u}_1$, that is $\vec{u}_3$ is scalar multiple of u_1, so it can be deleted from the spanning set of W. So the vectors $\vec{u}_1$ and $\vec{u}_2$ span W and are not scalar multiples of each other, hence are linearly independent. Thus, $\{\vec{u}_1, \vec{u}_2\}$ is a basis for W.

Example 1.34 : *Determine the dimensions of Nul A and Col A for the matrix.*

$$A = \begin{bmatrix} 1 & 3 & -4 & 2 & -1 & 6 \\ 0 & 0 & 1 & -3 & 7 & 0 \\ 0 & 0 & 0 & 1 & 4 & -3 \\ 0 & 0 & 0 & 0 & 0 & 0 \end{bmatrix}$$

Solution : We see that the matrix A is in row echelon form and it has three pivot columns, hence the dimension of the Col A is 3.

If we write down the corresponding system of equations of A, we have three basic variables and **three free variables**, so the dimension of the Nul A is 3.

1.7 The Rank of a Matrix

In this section, we introduce the important concept of rank related to matrix. The rows and columns of a matrix reveal several interesting and useful relationships. The significant amazing thing that we will see is that the dimensions of the column space and row space, though these are subspace of different vector spaces, are same. His common dimension is called the rank of the matrix. We first define the row space of a matrix.

Row Space :

Let A be an $m \times n$ matrix, then each row of A has n-entries and thus can be identified with a vector in $\mathbb{R}^n$. The set of all linear combinations of the row vectors of A is called a row space of A and is denoted by row A. In other words, the row A is the subspace of $\mathbb{R}^n$ generated by the row vectors of A. Thus, Row A is a subspace of $\mathbb{R}^n$.

Note :

1. The column space of A, Col A is a subspace of $\mathbb{R}^m$.

2. The rows of A can be identified with the columns of A^t, so we could also write Column A^t in place of Row A. That is, Row A and Col A^t are same.

Let us illustrate the above facts by an example.

$$\text{Let} \qquad A = \begin{bmatrix} 1 & 2 & 3 & 4 & 5 \\ 0 & 1 & -2 & -6 & 7 \\ 2 & 4 & -3 & 1 & 0 \\ 1 & 3 & 1 & -2 & 12 \end{bmatrix}$$

Now the rows of A can be considered as members of $\mathbb{R}^5$. So,

$$\vec{r_1} = (1, 2, 3, 4, 5)$$

$$\vec{r_2} = (0, 1, -2, -6, 7)$$

$$\vec{r_3} = (2, 4, -3, 1, 0)$$

$$\vec{r_4} = (1, 3, 1, -2, 12)$$

The row space of A is a subspace of $\mathbb{R}^5$ and is spanned by

$\{\vec{r_1}, \vec{r_2}, \vec{r_3}, \vec{r_4}\}$. That is Row $A = \text{Span } \{\vec{r_1}, \vec{r_2}, \vec{r_3}, \vec{r_4}\}$

If we knew some linear dependence relations among the rows of A, we could use the spanning set theorem to shrink the spanning set to a basis. In the above example, one can see that $\vec{r_1} + \vec{r_2} = \vec{r_4}$, that is $\vec{r_4}$ is linear combination of $\vec{r_1}$ and $\vec{r_2}$, hence can be eliminated from the spanning set. Thus, we have,

$$\text{Row } A = \text{Span } \{\vec{r_1}, \vec{r_2}, \vec{r_3}, \vec{r_4}\} = \text{Span } \{\vec{r_1}, \vec{r_2}, \vec{r_3}\}$$

We prove the following theorem which enables us to find the basis of for Row A.

Theorem 14 If two matrices A and B are row equivalent, then their row spaces are the same. If β is in echelon form, the non-zero rows of β form a basis for the row space of A as well as for that of B.

Proof : Remember that the row operations on a matrix A are the linear combinations of rows of A. Therefore, if B is obtained from A by row operations, then the rows of B are linear combinations of the rows of A. So it follows that any linear combination of rows of B is automatically a linear combination of the rows of A. Thus, the row space of B is contained in the row space of A. We know that the row operations are invertible, then we can obtain A from B by a sequence of row operations, hence the row space of is contained in row space of B. Hence the row spaces of A and B are same.

Now, if B is in echelon form, its non-zero rows are linearly independent because no non-zero row is a linear combination of the non-zero rows below it. Thus, the non-zero rows of B form a basis for the row space of B and that of A.   ■

Let us see the following example :

Illustrative Examples

Example 1.35 : *Find the bases for the row space, the column space and the null space of A, where*

$$A = \begin{bmatrix} 1 & 4 & 5 & 6 & 6 & 9 \\ 3 & -2 & 1 & 4 & 4 & -1 \\ -1 & 0 & -1 & -1 & -2 & -1 \\ 2 & 3 & 5 & 5 & 7 & 8 \end{bmatrix}$$

Solution : To find the bases for row space and column space of A, the row reduce A to an echelon form.

By applying $R_2 + (-3) R_1$, $R_3 + R_1$ and $R_4 + (-2) R_1$ on A we obtain,

$$A \sim \begin{bmatrix} 1 & 4 & 5 & 6 & 9 \\ 0 & -14 & -14 & -14 & -28 \\ 0 & 4 & 4 & 4 & 4 \\ 0 & -5 & -5 & -5 & -5 \end{bmatrix}$$

Next apply $\left(-\dfrac{1}{14}\right) R_2$ and then $R_3 + (-4) R_2$, $R_4 + (5) R_2$, we get

$$A \sim B = \begin{bmatrix} 1 & 4 & 5 & 6 & 9 \\ 0 & 1 & 1 & 1 & 2 \\ 0 & 0 & 0 & 0 & 0 \\ 0 & 0 & 0 & 0 & 0 \end{bmatrix}$$

Now, B is in row echelon form.

Then by theorem 13, the first two non-zero rows form a basis for the row space of A (as well as the basis for the row space of B). Thus, Basis for Row A is $\{(1, 4, 5, 6, 9), (0, 1, 1, 1, 2)\}$.

For the column space, we observe from B that first and second columns are pivot columns, hence the first and second columns of A form a basis for Col A :

$$\text{Basis for Col A is } \left\{ \begin{bmatrix} 1 \\ 3 \\ -1 \\ 2 \end{bmatrix}, \begin{bmatrix} 4 \\ -2 \\ 0 \\ 3 \end{bmatrix} \right\}.$$

Now to find the basis for Nul A, we need to row reduced echelon form. on B we apply $R_1 + (-4) R_2$ to get,

$$A \sim B \sim C = \begin{bmatrix} 1 & 0 & 1 & 2 & 1 \\ 0 & 1 & 1 & 1 & 2 \\ 0 & 0 & 0 & 0 & 0 \\ 0 & 0 & 0 & 0 & 0 \end{bmatrix}$$

Writing the corresponding system of linear equation of c, we get

$$x_1 + x_3 + 2x_4 + x_5 = 0$$
$$x_2 + x_3 + x_4 + 2x_5 = 0$$

OR
$$x_1 = -x_3 - 2x_4 - x_5$$

and
$$x_2 = -x_3 - x_n - 2x_5$$

x_1, x_2 are basic variables and x_3, x_4 and x_5 are free variables.

Thus, the general solution is given by,

$$\begin{bmatrix} x_1 \\ x_2 \\ x_3 \\ x_4 \\ x_5 \end{bmatrix} = \begin{bmatrix} -x_3 - 2x_4 - x_5 \\ -x_3 - x_4 - 2x_5 \\ x_3 \\ x_4 \\ x_5 \end{bmatrix}$$

Take $x_3 = r$, $x_4 = s$, $x_5 = t$ as parameters, $r, s, t \in \mathbb{R}$, we write the parametric solution as :

$$\begin{bmatrix} x_1 \\ x_2 \\ x_3 \\ x_4 \\ x_5 \end{bmatrix} = \begin{bmatrix} -r_3 - 2s_4 - t_5 \\ -r_3 - s_4 - 2t_5 \\ r \\ s \\ t \end{bmatrix} = r \begin{bmatrix} -1 \\ -1 \\ 1 \\ 0 \\ 0 \end{bmatrix} + s \begin{bmatrix} -2 \\ -1 \\ 0 \\ 1 \\ 0 \end{bmatrix} + t \begin{bmatrix} -1 \\ -2 \\ 0 \\ 0 \\ 1 \end{bmatrix}$$

Thus, the basis for the null space of A is :

$$\text{Basis for Nul A : } \left\{ \begin{bmatrix} -1 \\ -1 \\ 1 \\ 0 \\ 0 \end{bmatrix}, \begin{bmatrix} -2 \\ -1 \\ 0 \\ 1 \\ 0 \end{bmatrix}, \begin{bmatrix} -1 \\ -2 \\ 0 \\ 0 \\ 1 \end{bmatrix} \right\}$$

Note : In the above example (35), A is 4×5 matrix, dim Row A = 2, dim Col A = 2 and dim Nul A = 3. We see that

$$\text{Row} = A = \text{Col } A = 2 \quad \text{and}$$

$$\text{dim Row A (or dim Col A)} + \text{dim Nul A} = 5$$

$$= 2 + 3 = 5 \text{ the number of columns of A.}$$

Definition : The **rank** of a matrix A is the dimension of column space of A.

Note : We know that Row A is the same as Col A^t, the dimension of the row space of A is the rank of A^t.

Definition : The dimension of the null space of A is called the **nullity** of A.

$\boxed{\textbf{Theorem 15}}$ **(The Rank Theorem) :**

The dimensions of column space and row space of an $m \times n$ matrix A are equal. This common dimension, the rank of A, also equals the number of pivot positions in A and satisfies the equation.

$$\text{Rank A} + \text{dim Nul A} = n.$$

Proof : We know that the rank A is the number of pivot columns in A. Equivalently, rank A is the number of pivot positions in an echelon form B of A. Since, B has a non-zero row for each pivot, and since these rows form a basis for the row space of A, the rank of A is also the dimension of the row space.

Next, we have seen that the dimension of Nul A equals the number free variables in the equation $A\vec{x} = \vec{0}$. In other way, the dimension of Nul A is the number of columns of A that are not pivot columns. Thus,

$$\left\{\begin{array}{c}\text{The number of} \\ \text{pivot columns}\end{array}\right\} + \left\{\begin{array}{c}\text{The number of} \\ \text{non-pivot column}\end{array}\right\} = \left\{\begin{array}{c}\text{The number of} \\ \text{columns}\end{array}\right\}$$

That is rank A + dim Nul A = n.   ∎

Example 1.36 : *(a) If A is 5×8 matrix and dim Nul A = 4, then what is rank of A?*

(b) If a matrix A is of size 4×7, is it possible that dim Nul A = 2 ?

Solution : (a) By the Rank Theorem, since A has 8-columns.

$$\text{Rank A} + \text{dim Nul A} = 8.$$

$$\text{Rank A} + 4 = 8.$$

Therefore, Rank A = 4.

(b) Number if dim Nul A = 2, then by the Rank Theorem, rank A must be 5, since A has 7-columns. But the columns of A are members of $\mathbb{R}^4$, so dimension of Col A cannot exceed 4, that is rank A cannot exceed 4.

Rank and the Invertible Matrix Theorem :

One more theorem providing the equivalence statements relating rank of a matrix and invertibility.

$\boxed{\textbf{Theorem 16}}$ **The Invertible Matrix Theorem**

Let A be an n × n matrix. Then the following statements are equivalent to the statement that A is an invertible matrix.

(m) The columns of A form a basis of $\mathbb{R}^n$.

(n) Col A = $\mathbb{R}^n$.

(o) Dim Col A = n.

(p) Rank A = n.

(q) Nul A = $\{\vec{0}\}$.

(r) Dim Nul A = 0.

Proof : If A is invertible then we know that the columns of A are linearly independent and span $\mathbb{R}^n$, hence the columns of A form a basis and vice versa.

We know that (g) if A is invertible then the equation $A\vec{x} = \vec{b}$ has at least one solution for each $\vec{b}$ in $\mathbb{R}^2$? This shows that each vector $\vec{b}$ in $\mathbb{R}^n$ is a linear combination of columns of A, that is Col A = $\mathbb{R}^n$. (n) ⇒ (o) ⇒ (p) follow from the definitions of dimension and rank. Next, if rank A = n, the number of columns of A, then dim Nul A = 0, by the Rank Theorem; so Nul A = $\{\vec{0}\}$.

Thus, (p) ⇒ (r) ⇒ (q). Also (q) implies that (d) the equation $A\vec{x} = \vec{0}$ has only trivial solution, which means the statements (g) and (d) are equivalent and both are equivalent to the statement (a) A is invertible.   ∎

Example 1.37 : *Let A and B be the matrices given below.*

$$A = \begin{bmatrix} 2 & -3 & 6 & 2 & 5 \\ -2 & 3 & -3 & -3 & -4 \\ 4 & -6 & 9 & 5 & 9 \\ -2 & 3 & 3 & -4 & 1 \end{bmatrix} \quad B = \begin{bmatrix} 2 & -3 & 6 & 2 & 5 \\ 0 & 0 & 3 & -1 & 1 \\ 0 & 0 & 0 & 1 & 3 \\ 0 & 0 & 0 & 0 & 0 \end{bmatrix}$$

Suppose A is row equivalent to B.

(a) Just by mere observation state rank A and dim Nul A.

(b) Then find bases for Col A, Row A and Nul A.

Solution : (a) Since, there are three pivot columns in B, also dim Col B = 3 hence dim Col A = 3, and Rank A = 3. By the Rank Theorem as there are 5 columns in A, dim Nul A = 5 − 3 = 2.

(b) As in matrix B, the columns 1, 3 and 4 are pivot columns, the columns 1, 3 and 4 of A form a basis for Col A, that is basis for Col A is

$$\left\{ \begin{bmatrix} 2 \\ -2 \\ 4 \\ -2 \end{bmatrix}, \begin{bmatrix} 6 \\ -3 \\ 9 \\ 3 \end{bmatrix}, \begin{bmatrix} 2 \\ -3 \\ 5 \\ -4 \end{bmatrix} \right\}$$

By the theorem 13, the first three rows of B form a basis for the row space of A as well as for the row space of B. Thus,

Basis for Row A = {(2, −3, 6, 2, 5), (−2, 3, −3, −3, −4), (4, −6, 9, 5, 9)}

Now to find the basis for Nul A, we row reduce the matrix B, so that C is row equivalent to B and A.

We apply $R_1 + (-2) R_3$, $R_2 + R_3$

$$B \sim \begin{bmatrix} 2 & -3 & 6 & 0 & -1 \\ 0 & 0 & 3 & 0 & 4 \\ 0 & 0 & 0 & 1 & 3 \\ 0 & 0 & 0 & 0 & 0 \end{bmatrix}$$

Now apply $R_1 + (-2) R_2$ and then $\frac{1}{3} R_2$, we obtain

$$A \sim B \sim C = \begin{bmatrix} 2 & -3 & 0 & 0 & -9 \\ 0 & 0 & 1 & 0 & 4/3 \\ 0 & 0 & 0 & 1 & 3 \\ 0 & 0 & 0 & 0 & 0 \end{bmatrix}$$

Now writing down the system $c\vec{x} = \vec{0}$, we get

$$2x_1 - 3x_2 - 9x_5 = 0$$

$$x_3 + \frac{4}{3} x_5 = 0$$

$$x_4 + 3x_5 = 0$$

There are three basic variables x_1, x_3 and x_4 and two free variables namely x_2 and x_5. Thus, we write

$$x_1 = 3x_2 + 9x_5$$

$$x_3 = -\frac{4}{3} x_5$$

$$x_4 = -3x_5$$

Thus, the parametric solution of the system $(\vec{x} = \vec{0})$ is given by

$$\begin{bmatrix} x_1 \\ x_2 \\ x_3 \\ x_4 \\ x_5 \end{bmatrix} = \begin{bmatrix} 3r + gs \\ r \\ -4/3\ s \\ -3s \\ s \end{bmatrix} \quad x_2 = r,\ x_5 = s,\ r,\ s \in \mathbb{R}$$

$$= \begin{bmatrix} 3 \\ 1 \\ 0 \\ 0 \\ 0 \end{bmatrix} + s \begin{bmatrix} 9 \\ 0 \\ -4/3 \\ -3 \\ 1 \end{bmatrix}$$

The basis for Null C, and hence the basis for Nul A is

$$\left\{ \begin{bmatrix} 3 \\ 1 \\ 0 \\ 0 \\ 0 \end{bmatrix}, \begin{bmatrix} 9 \\ 0 \\ -4/3 \\ -3 \\ 1 \end{bmatrix} \right\}$$

Example 1.38 : *If A is 7 × 5 matrix, what is the largest possible of rank A ? If A is 5 × 7 matrix, what is the largest possible rank A ? Explain.*

Solution : Since, rank A is the number of pivot largest possible rank A is 5. In second case though there are 7 columns, but only 5 rows, so the row-reduced echelon form of A could have maximum 5 pivot columns.

Think Over It

1. Let W be the set of all vectors of the form

$$\begin{bmatrix} s + 3t \\ s - t \\ 2s - t \\ 4t \end{bmatrix}$$

   Is every vector in W a linear combination of two particular vectors ?

   If your answer is yes, which are the two vectors ?

2. Given an m × n matrix A, an element in Col A has the form $A\vec{x}$ for some $\vec{x}$ in $\mathbb{R}^n$. Why zero vector is in Col A ?

3. Find a basis for the subspace spanned by the set

   {sin t, sin 2t, sin t cos t}

   in a vector space of all real valued functions.

4. Let S be a finite set in a vector space V with property that every vector in V has a unique representation as a linear combination of elements in S. What can you say about the set S ?

5. Is space P of all polynomials an infinite dimensional space ?

6. If A is a 6 × 8 matrix, what is the smallest possible dimension of Nul A ?

Summary

1. Vector spaces, subspaces their examples.

2. Null space of a matrix A and its relation with the system $A\vec{x} = \vec{0}$. Basis for Null space. Column space of matrix, spanning sets for column space Kernel and Range of a linear transformation.

3. Linear dependent and independent sets, linear span, the spanning set theorem bases for null space and column space.

4. Co-ordinate systems in vector spaces Unique Representation Theorem, co-ordinates in $\mathbb{R}^n$. The co-ordinate mapping.

5. The dimension of a vector space. Subspaces of finite dimensional spaces. The Basis Theorem. The dimension of Null space and Column space.

6. The Rank of a matrix. Row space of a matrix. The rank theorem, rank and invertible matrix theorem.

Exercise

[A] Say True or False : Justify !

1. $\mathbb{R}^3$ is a subspace of $\mathbb{R}^5$.

2. A subspace is also a vector space.

3. A null space of an $m \times n$ matrix is a subspace of $\mathbb{R}^m$.

4. Nul A is the kernel and mapping $\vec{x} \to A\vec{x}$.

5. A single non-zero vector in a vector space is linearly dependent.

6. A basis is a linearly independent set that is as large as possible.

7. The correspondence $[\vec{x}]_\beta - \vec{x}$ is called the co-ordinate mapping.

8. The vector spaces P_5 and $\mathbb{R}^5$ are isomorphic.

9. A plane in $\mathbb{R}^3$ is a two-dimensional subspace of $\mathbb{R}^3$.

10. The number of variables in the equation $A\vec{x} = \vec{0}$ equals the dimension of Nul A.

11. If B is any echelon form of A, and if B has three non-zero rows, then the first three rows of A form a basis for Row A.

12. The dimension of the null space of A is the number of columns of A that are not pivot columns.

[B] Multiple Choice Questions : Choose the Correct Alternative

1. The set $S = \{\vec{u}_1, \vec{u}_2, ..., \vec{u}_p\}$ spans a vector space V, then ......
   (a) S is a basis for V
   (b) dim V = p
   (c) S need not be a basis for V
   (d) None of these

2. In a vector space V, $\vec{u} \in V$ be any vector and α is any scalar such that $\alpha\vec{u} = 0$, then ......

   (a) $\alpha \neq 0$
   (b) $\vec{u} \neq \vec{0}$

   (c) either $\alpha = 0$ or $\vec{u} = \vec{0}$
   (d) none of these

3. If A is m × n matrix, m ≠ n, then ......
    (a) Nul A is subspace of $\mathbb{R}^n$.
    (b) Nul A and Col A are subspaces of $\mathbb{R}^m$.
    (c) Col A is subspace of $\mathbb{R}^n$.
    (d) None of these

4. If A is m × n matrix, and $A\vec{x} = \vec{b}$ is consistent for each $\vec{b}$ in $\mathbb{R}^m$, then ......
    (a) Col A = $\mathbb{R}^n$
    (b) Col A = $\mathbb{R}^m$
    (c) Col A is proper subspace of $\mathbb{R}^m$
    (d) None of these

5. Pivot columns of a matrix form a basis for ......
    (a) Row A   (b) Col A
    (c) Nul A   (d) None of these

6. The columns of an invertible n × n matrix form a basis for ......
    (a) Col A   (b) Row A
    (c) Nul A   (d) (a) as well as (b) but not (c)

7. If dim V = n, then the co-ordinate map $\vec{x} \to [\vec{x}]_\beta$ is ......
    (a) not one-to-one   (b) not onto
    (c) an isomorphism   (d) none of these

8. Let B = $\{\vec{u}_1, \vec{u}_2, ..., \vec{u}_n\}$ is a basis for a vector space V, then $[\vec{x}]_\beta$ is ......
    (a) a vector in $\mathbb{R}^n$ (b) a vector in $\mathbb{R}^m$
    (c) a vector in V   (d) none of these

9. Let V be an n-dimensional vector space, then ......
    (a) any linearly independent set of exactly n-vectors in V is basis for V.
    (b) any set of exactly n-vectors that spans V is a basis for V.
    (c) any set of n-vectors is basis for V.
    (d) (a) and (b) are both true.

10. A dimension of a vector space P_4 is ......
    (a) 4   (b) 5
    (c) 6   (d) None of these

11. If A is 7×9 matrix, has two-dimensional null space, then the rank A is ......

    (a) 7   (b) 9

    (c) 6   (d) None of these

12. Row space of A is ......

    (a) same as column space of A^t.   (b) same as column space of A.

    (c) (a) is true but not (b).   (d) none of these

[C] Theory Questions :

1. If $\{\vec{u}_1, \vec{u}_2, ..., \vec{u}_k\}$ is a set vectors in a vector space V, then prove that span $\{\vec{u}_1, ..., \vec{u}_2, ..., \vec{u}_k\}$ is a subspace of V.

2. Define null space of an $m \times n$ matrix A. Prove that Nul A is subspace of $\mathbb{R}^n$.

OR

If A is $m \times n$ matrix, then prove that the set of all solutions of the system $A\vec{x} = \vec{0}$ of m homogeneous linear equations in n-unknowns is a subspace of $\mathbb{R}^n$.

3. Prove that the column space of an $m \times n$ matrix A is a subspace of $\mathbb{R}^m$.

4. Prove that an indexed set $\{\vec{u}_1, ..., \vec{u}_k\}$ with $\vec{u}_1 \neq \vec{0}$ and $k \geq 2$, is linearly dependent if and only if some $\vec{u}_j$ (with $j > 1$) is a linear combination of the preceding vectors $\vec{u}_1, ..., \vec{u}_{j-1}$.

5. State and prove the spanning set theorem.

6. Prove that the pivot columns of a matrix A form a basis for Col A.

7. State and prove the Unique Representation Theorem.

8. Let $\beta = \{\vec{u}_1, \vec{u}_2, ..., \vec{u}_n\}$ be a basis for a vector space V. Then prove that the co-ordinate map $\phi : V \to \mathbb{R}^n$ given by $\phi(\vec{u}) = [\vec{u}]_\beta$ is a one-to-one linear transformation.

9. Let $\beta = \{\vec{u}_1, \vec{u}_2, ..., \vec{u}_n\}$ be a basis for a vector space V. Then prove that any set containing more than n-vectors in V must be linearly dependent.

10. Prove that in a finite dimensional vector space V, every basis of V must have the same number of vectors.

11. Let H be a subspace of a finite dimensional vector space V. Prove that any linearly independent set in H can be enlarged, if necessary, to a basis for H. Also, show that H is finite dimensional and dim H $\leq$ dim V.

12. State and prove the Basis Theorem.

13. Prove that row spaces of two row equivalent matrices are same. Further, prove that if B is in row echelon form and A is row equivalent to B, then non-zero rows of B form a basis for the row space of A as well as for that of B.

14. State and prove the Rank Theorem.

[D] Numerical Problems :

1. Let V be the set of all real valued function defined on a subset D of real numbers. Show that V is a vector space under addition of functions and scalar multiplication of functions.

2. Let V be a vector space and $\vec{u}_1$, $\vec{u}_2$ are vectors in V, then show that H = Span $\{\vec{u}_1, \vec{u}_2\}$ is a subspace of V.

3. Let H be the set of all vectors of the form $(a + 2b, 2a, 3b, b - a)$, $a, b \in \mathbb{R}$. Show that H is a subspace of $\mathbb{R}^4$.

4. Show that the set H of vectors in $\mathbb{R}^3$ of the form $(2a, b - a, 1 + a + b)$ is not a subspace of $\mathbb{R}^3$.

5. Let $W = \left\{ \begin{bmatrix} x \\ y \end{bmatrix} : xy \geq 0 \right\}$. Find two specific vectors $\vec{u}$ and $\vec{v}$ in W such that $\vec{u} + \vec{v}$ is not in W. Is W a subspace of $\mathbb{R}^2$?

6. Let $\vec{u}_1 = \begin{bmatrix} 1 \\ 0 \\ -1 \end{bmatrix}$ $\vec{u}_2 = \begin{bmatrix} 2 \\ 1 \\ 3 \end{bmatrix}$, $\vec{u}\vec{u} = \begin{bmatrix} 4 \\ 2 \\ 6 \end{bmatrix}$ and $\vec{w} = \begin{bmatrix} 3 \\ 1 \\ 2 \end{bmatrix}$.

   (a) Is $\vec{w}$ in $\{\vec{u}_1, \vec{u}_2, \vec{u}_3\}$? How many vectors are there in $\{\vec{u}_1, \vec{u}_2, \vec{u}_3\}$?

   (b) How many vectors are in Span $\{\vec{u}_1, \vec{u}_2, \vec{u}_3\}$?

   (c) Is $\vec{w}$ in the subspace spanned by $\{\vec{u}_1, \vec{u}_2, \vec{u}_3\}$? Why ?

7. Find the spanning set for the null space of the matrix :

$$A = \begin{bmatrix} -3 & 6 & -1 & 1 & -7 \\ 1 & -2 & 2 & 3 & -1 \\ 2 & -4 & 5 & 8 & -4 \end{bmatrix}.$$

8. Find a matrix A such that $W = \text{Col } A$;

$$W = \left\{ \begin{bmatrix} a + b + c \\ 2a - c \\ b + c \end{bmatrix} : a, b, c \in \mathbb{R} \right\}$$

9. Let $A = \begin{bmatrix} 2 & 4 & -2 & 1 \\ -2 & -5 & 7 & 3 \\ 3 & 7 & -8 & 6 \end{bmatrix}$

   (a) Find a non-zero vector in Col A.

   (b) Find a non-zero vector in Nul A.

   (c) Show that $\vec{u} = \begin{bmatrix} 3 \\ -2 \\ -1 \\ 0 \end{bmatrix}$ is in Nul A.

10. Let $A = \begin{bmatrix} -8 & -2 & -9 \\ 6 & 4 & 8 \\ 4 & 0 & 4 \end{bmatrix}$ and $\vec{w} = \begin{bmatrix} 2 \\ 1 \\ -2 \end{bmatrix}$. Determine if $\vec{w}$ is in Col A. Is $\vec{w}$ in Nul A ?

11. Let $S = \left\{ \begin{bmatrix} 1 \\ 0 \\ -2 \end{bmatrix}, \begin{bmatrix} 3 \\ 2 \\ -4 \end{bmatrix}, \begin{bmatrix} -3 \\ -5 \\ 1 \end{bmatrix} \right\}$ and $T = \left\{ \begin{bmatrix} 1 \\ -3 \\ 0 \end{bmatrix}, \begin{bmatrix} -2 \\ 9 \\ 0 \end{bmatrix}, \begin{bmatrix} 0 \\ 0 \\ 0 \end{bmatrix}, \begin{bmatrix} 0 \\ -3 \\ 5 \end{bmatrix} \right\}.$

   Determine where S and T are basis for $\mathbb{R}^3$. Of the sets S and T which is not basis, determine which one is linearly independent and which one Spans $\mathbb{R}^3$.

12. Let $\vec{u}_1 = \begin{bmatrix} 4 \\ -3 \\ 7 \end{bmatrix}, \vec{u}_2 = \begin{bmatrix} 1 \\ 9 \\ -2 \end{bmatrix}, \vec{u}_3 = \begin{bmatrix} 7 \\ 11 \\ 6 \end{bmatrix}$, and let

   $H = \text{Span } \{\vec{u}_1, \vec{u}_2, \vec{u}_3\}.$

   Observe that $4\vec{u}_1 + 5\vec{u}_2 + 3\vec{u}_3 = \vec{0}$. Using this observations determine the basis for H.

13. Let $p_1(x) = 1 + x^2$ and $p_2(x) = 1 - x^2$. Is $\{p_1, p_2\}$ a linearly independent set in P_3 ? Justify.

14. Let $\beta = \left\{ \begin{bmatrix} 1 \\ 0 \end{bmatrix}, \begin{bmatrix} 1 \\ 2 \end{bmatrix} \right\}$ be a basis for $\mathbb{R}^2$.

    (a)   Find $[\vec{x}]_\beta$, where $\vec{x} = \begin{bmatrix} -1 \\ 1 \end{bmatrix}$

    (b)   Find $\vec{x}$, such that $[\vec{x}] = \begin{bmatrix} -2 \\ 3 \end{bmatrix}$.

15. Let $p_1(x) = 1 + 2x^2$, $p_2(x) = 4 + x + 5x^2$, $p_3(x) = 3 + 2t$.

    (a)   Find co-ordinate vectors of each polynomial relative to the standard basis for P_2.

    (b)   Use these co-ordinate vector to show that $\{p_1, p_2, p_3\}$ is linearly dependent.

16. Let $\vec{u}_1 = \begin{bmatrix} 1 \\ 0 \\ 0 \end{bmatrix}$, $\vec{u}_2 = \begin{bmatrix} -3 \\ 4 \\ 0 \end{bmatrix}$, $\vec{u}_3 = \begin{bmatrix} 3 \\ -6 \\ 3 \end{bmatrix}$, and $\vec{x} = \begin{bmatrix} -8 \\ 2 \\ 3 \end{bmatrix}$.

    (a)   Show that $B = \{\vec{u}_1, \vec{u}_2, \vec{u}_3\}$ is a basis of $\mathbb{R}^3$.

    (b)   Find the change of co-ordinate matrix from B to the standard basis.

    (c)   Write relation that relates $\vec{x}$ in $\mathbb{R}^3$ to $[\vec{x}]_b$.

    (d)   Find $[\vec{x}]_B$, for $\vec{x}$ given above.

17. Write down the standard bases for $\mathbb{R}^2$, $\mathbb{R}^3$, $\mathbb{R}^n$, P_2, P_3, P_n.

18. Find the dimension of the subspace

$$H = \left\{ \begin{bmatrix} a - 3b + 6c \\ 5a + 4d \\ b - 2c - d \\ 5d \end{bmatrix} : a, b, c, d \in \mathbb{R} \right\}$$

19. Find the dimension of the null space and column space of

$$A = \begin{bmatrix} -3 & 6 & -1 & 1 & -7 \\ 1 & -2 & 2 & 3 & -1 \\ 2 & -4 & 5 & 8 & -4 \end{bmatrix}$$

20. Find the dimension of the subspace spanned by

$$\left\{ \begin{bmatrix} 1 \\ 0 \\ 2 \end{bmatrix}, \begin{bmatrix} 3 \\ 1 \\ 1 \end{bmatrix}, \begin{bmatrix} 9 \\ 4 \\ -2 \end{bmatrix}, \begin{bmatrix} -7 \\ -3 \\ 1 \end{bmatrix} \right\}$$

21. Find the bases for the row space, the column space and the null space of the matrix.

$$A = \begin{bmatrix} -2 & -5 & 8 & 0 & -17 \\ 1 & 3 & -5 & 1 & 5 \\ 3 & 11 & -19 & 7 & 1 \\ 1 & 7 & -13 & 5 & -3 \end{bmatrix}$$

22. Assume that the matrix A is row-equivalent to B. Without calculations find rank A and dim Nul A. Then find bases for Col A, Row A and Nul A, where,

$$A = \begin{bmatrix} 1 & -4 & 9 & -7 \\ -1 & 2 & -4 & 1 \\ 5 & -6 & 10 & 7 \end{bmatrix} \text{ and } B = \begin{bmatrix} 1 & 0 & -1 & 5 \\ 0 & -2 & 5 & -6 \\ 0 & 0 & 0 & 0 \end{bmatrix}$$

23. Suppose a 4×7 matrix A has four pivot columns. Is Col A = $\mathbb{R}^4$? Is Nul A = $\mathbb{R}^3$.

24. If A is a 4×3 matrix, what is the largest possible dimension of row space of A ? If A is 3×4 matrix, what is the largest possible dimension of A ? Justify.

25. If the null space of an 8×5 matrix A is 2-dimensional, what is dimension of the row space of A ?

Answers

[A] (1) False (2) True (3) False (4) True (5) False (6) True

    (7) False (8) False

    (9) False, only planes passing through origin are subspace

    (10) False (11) True (12) True

[B] (1) - (c) (2) - (c) (3) - (a) (4) - (b) (5) - (b) (6) - (d) (7) - (c)

    (8) - (a) (9) - (d) (10) - (b) (11) - (a) (12) - (c)

[D] 3. **Hint :** Every vector in H is a linear combination of the vectors

$$\vec{u}_1 = \begin{bmatrix} 1 \\ 2 \\ 0 \\ -1 \end{bmatrix} \text{ and } \vec{u}_2 = \begin{bmatrix} 2 \\ 0 \\ 3 \\ 1 \end{bmatrix}.$$

4. The zero is not a member of H.

5. $\vec{u} = \begin{bmatrix} -2 \\ -3 \end{bmatrix}$ and $\vec{v} = \begin{bmatrix} 0 \\ 5 \end{bmatrix}$, $\vec{u} + \vec{v} = \begin{bmatrix} -2 \\ 2 \end{bmatrix}$ is not in W.

W is not subspace.

6. (a) No, there are only 3 vectors.

(b) There are infinitely many vectors in Span $\{\vec{u}_1, \vec{u}_2, \vec{u}_3\}$.

(c) W is in Span $\{\vec{u}_1, \vec{u}_2, \vec{u}_3\}$. $\vec{w}$ is linear combination of $\vec{u}_1, \vec{u}_2, \vec{u}_3$. $\vec{w} = \vec{u}_1 - \vec{u}_2 + \vec{u}_3$.

7. $\left\{ \begin{bmatrix} 2 \\ 1 \\ 0 \\ 0 \\ 0 \end{bmatrix}, \begin{bmatrix} 1 \\ 0 \\ -2 \\ 1 \\ 0 \end{bmatrix}, \begin{bmatrix} -3 \\ 0 \\ 2 \\ 0 \\ 1 \end{bmatrix} \right\}$ Spans Nul A.

8. $A = \begin{bmatrix} 1 & 1 & 1 \\ 2 & 0 & -1 \\ 0 & 1 & 1 \end{bmatrix}.$

9. (a) $\begin{bmatrix} 2 \\ -2 \\ 3 \end{bmatrix}$ is in Col A. In fact, each column vector in A is in Col A.

(b) $\begin{bmatrix} -9 \\ 5 \\ 1 \\ 0 \end{bmatrix}$ is in Nul A.

(c) $A\vec{u} = \vec{0}$, verify.

10. $\vec{w}$ is in Col A. $\vec{w}$ in Nul A.

11. S is not a basis, it is linearly dependent and do not span $\mathbb{R}^3$.

T is not basis, since there are four vectors in T. T is linearly dependent as zero vector belongs to T.

T spans $\mathbb{R}^3$, since the matrix formed by these vectors has 3-pivot columns.

12. One of the vectors $\vec{u}_1$, $\vec{u}_2$, $\vec{u}_3$ is a linear combination of the remaining two vectors, and no two vectors are scalar multiples of each other. Hence, any two of the vectors $\vec{u}_1$, $\vec{u}_2$, $\vec{u}_3$ forms a basis for H.

13. $\{P_1, P_2\}$ is linearly independent.

14. (a) $[\vec{x}]_B = \begin{bmatrix} \dfrac{-1}{2} \\ \dfrac{1}{2} \end{bmatrix}$ (b) $\vec{x} = \begin{bmatrix} 1 \\ 6 \end{bmatrix}$

15. (a) $[P_1]_\beta = (1, 0, 2)$, $[P_2]_\beta = (4, 1, 5)$, $[P_3]_B = (3, 2, 0)$

    (b) The matrix of co-ordinate vectors in echelon form shows that P_1, P_2, P_3 are linearly dependent. Infact,
$$3 + 2x = 2(4 + x + 5x^2) - 5(1 + 2x^2)$$

16. (a) β is linearly independent hence is a basis

    (b) $P_\beta = \begin{bmatrix} 1 & -3 & 3 \\ 0 & 4 & -6 \\ 0 & 0 & 3 \end{bmatrix}$ (c) $\vec{x} = P_\beta[\vec{x}]_\beta$

    (d) $[\vec{x}]_\beta = \begin{bmatrix} -5 \\ 2 \\ 1 \end{bmatrix}$

17. Standard basis of :

    (i) $\mathbb{R}^2$ is $\left\{ \begin{bmatrix} 1 \\ 0 \end{bmatrix}, \begin{bmatrix} 0 \\ 1 \end{bmatrix} \right\}$

    (ii) $\mathbb{R}^3$ is $\left\{ \begin{bmatrix} 1 \\ 0 \\ 0 \end{bmatrix}, \begin{bmatrix} 0 \\ 1 \\ 0 \end{bmatrix}, \begin{bmatrix} 0 \\ 0 \\ 1 \end{bmatrix} \right\}$

    (iii) $\mathbb{R}^n$ is $\left\{ \begin{bmatrix} 1 \\ 0 \\ \vdots \\ 0 \end{bmatrix}, \begin{bmatrix} 0 \\ 1 \\ 0 \\ \vdots \\ 0 \end{bmatrix} \cdots \begin{bmatrix} 0 \\ 0 \\ \vdots \\ 1 \end{bmatrix} \right\}$

    Standard basis for :

    (i) P_2 is $\{1, t, t^2\}$ (ii) P_3 is $\{1, t, t^2, t^2\}$

    (iii) P_n is $\{1, t, \ldots, t^n\}$

18. $\left\{ \begin{bmatrix} 1 \\ 5 \\ 0 \\ 0 \end{bmatrix}, \begin{bmatrix} -3 \\ 0 \\ 1 \\ 0 \end{bmatrix}, \begin{bmatrix} 0 \\ 4 \\ -1 \\ 5 \end{bmatrix} \right\}$

19. dim Nul A = 3, dim Col A = 2.

20. 2

21. Basis for row space of A :

$\{(1, 3, -5, 1, 5), \{0, -2, 2, -7), (0, 0, 0, -4, 20)\}$

Basis for the column space of A : $\left\{ \begin{bmatrix} -2 \\ 1 \\ 3 \\ 1 \end{bmatrix}, \begin{bmatrix} -5 \\ 3 \\ 11 \\ 7 \end{bmatrix}, \begin{bmatrix} 0 \\ 1 \\ 7 \\ 5 \end{bmatrix} \right\}$

The basis for Nul A $= \left\{ \begin{bmatrix} -1 \\ 2 \\ 1 \\ 0 \\ 0 \end{bmatrix}, \begin{bmatrix} -1 \\ -3 \\ 0 \\ 5 \\ 1 \end{bmatrix} \right\}$

22. Rank A = 2, dim Nul A = 2

Basis for Col A : $\left\{ \begin{bmatrix} 1 \\ -1 \\ 5 \end{bmatrix}, \begin{bmatrix} -4 \\ 2 \\ -6 \end{bmatrix} \right\}$

Basis for Row A : $\{(1, 0, -1, 5), (0, -2, 5, -6)\}$

Basis for Nul A : $\left\{ \begin{bmatrix} 1 \\ 5/2 \\ 1 \\ 0 \end{bmatrix}, \begin{bmatrix} -5 \\ -3 \\ 0 \\ 1 \end{bmatrix} \right\}.$

23. Yes, Col A $= \mathbb{R}^4$. No, Nul A not $\mathbb{R}^3$. dim Nul A = 3, however Nul A is subspace of $\mathbb{R}^7$.

24. Since, 4 rows and three columns in A, largest possible dimension of row space of A is 3; as there is possibility of maximum three pivot columns.

Similar for 3×4, the largest possible dimension of A is 3.

25. Since, dim Col A = dim Row A = rank A. Therefore, by the Rank Theorem, Rank A + dim Nul A = 5, the number of columns in A.

$\therefore$ Rank = 5 – 2 $\because$ dim Nul A = 2

            = 3

Hence, dim Row A = 3

☞ ☞ ☞

Chapter 2...
Eigenvalues and Eigenvectors

Arthur Lee Samuel was an American scientist in the field of Computer gaming and artificial intelligence. He coined the term "machine learning" in 1959. He was also a senior member in the Tex community which were used in mathematical typographical system.

Arthur Lee Samuel

2.1 Introduction

In this section, we study the transformation $\vec{x} \rightarrow A\vec{x}$ in different perspective, namely $A\vec{x}$ is a scalar multiple of $\vec{x}$. This defines two concepts/terms **eigenvalue** and **eigenvector** of a matrix. The matrix used is always a square matrix. These concepts are useful throughout pure and applied mathematics. Eigenvalues are also used to study differential equations and continuous dynamical systems, they provide critical information in engineering design, and arise naturally in fields such as physics and chemistry.

Given a square matrix, we will find eigenvalues and then corresponding Eigenvectors. The **eigenspace** of eigenvalue is obtained using null space of the matrix $(\lambda I - A)$. We will study the characteristics of the eigenvectors corresponding to the eigenvalues of a matrix. The **characteristic** equation of a matrix is obtained by determinant expansion. We will study the diagonalization of a matrix related to eigenvalues and eigenvectors. In the last section, we will study the eigenvectors of linear transformations.

2.2 Eigenvectors and Eigenvalues

As we know that the transformation $\vec{x} \to A\vec{x}$ moves vectors $\vec{x}$ in variety of directions, but there some special vectors on which the action of A is quite simple.

Let us see this in the following example.

Example 2.1 : *Let* $A = \begin{bmatrix} 3 & 0 \\ 4 & -1 \end{bmatrix}$, $\vec{u} = \begin{bmatrix} 1 \\ 1 \end{bmatrix}$ *and* $\vec{v} = \begin{bmatrix} 1 \\ -1 \end{bmatrix}$. *What are the images of* $\vec{u}$ *and* $\vec{v}$ *under the transformation* $\vec{x} \to A\vec{x}$? *Represent all this by diagram. Comment on your observations.*

Solution : We have, $A\vec{u} = \begin{bmatrix} 3 & 0 \\ 4 & -1 \end{bmatrix} \begin{bmatrix} 1 \\ 1 \end{bmatrix} = \begin{bmatrix} 3 \\ 3 \end{bmatrix} = 3 \begin{bmatrix} 1 \\ 1 \end{bmatrix} = 3\vec{u}$

And. $\qquad A\vec{v} = \begin{bmatrix} 3 & 0 \\ 4 & -1 \end{bmatrix} \begin{bmatrix} 1 \\ -1 \end{bmatrix} = \begin{bmatrix} 3 \\ 5 \end{bmatrix}$

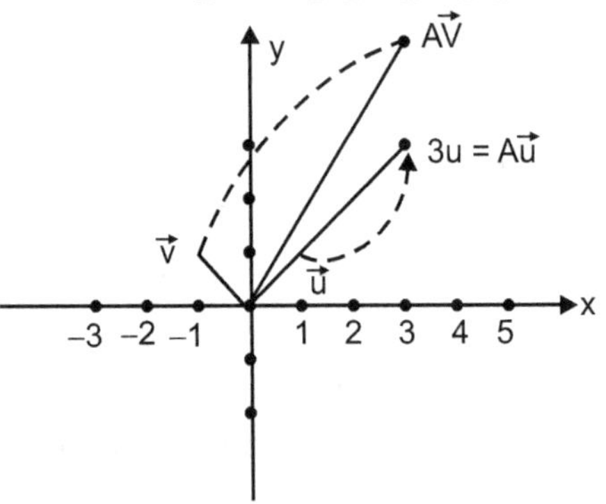

Fig. 2.1

The image of $\vec{u}$ is $3\vec{u}$ and image of $\vec{v}$ is $\begin{bmatrix} 3 \\ 5 \end{bmatrix}$.

We observe that image of $\vec{u}$ under the transformation is 3-times $\vec{u}$, where as image of $\vec{v}$ is $\begin{bmatrix} 3 \\ 5 \end{bmatrix}$ which is not scalar multiple of $\vec{v}$. We see that A stretches only $\vec{u}$. 3 is called eigenvalue of A and $\begin{bmatrix} 1 \\ 1 \end{bmatrix} = \vec{u}$ is called eigenvector of A corresponding to eigenvalue 3. We define formally these concepts and the method to find eigenvalues and eigenvectors of a given matrix.

Definition : Let A be an $n \times n$ matrix. An **eigenvector** of A is a non-zero vector $\vec{x}$ in $\mathbb{R}^n$ such that $A\vec{x} = \lambda\vec{x}$ for some scalar λ. A scalar λ is called an **eigenvalue** of A if there is a non-trivial solution $\vec{x}$ of the equation $A\vec{x} = \lambda\vec{x}$, such $\vec{x}$ is called an eigenvector of A corresponding to λ.

Note :

1. An eigenvector must be non-zero, by definition.

2. An eigenvalue may be zero. We will see this case when a number 0 is an eigenvalue in further work out.

3. To determine if a given vector is an eigenvector of the matrix is easy. It is also easy to decide if a specified scalar is an eigenvalue.

Example 2.2 : *Let* $A = \begin{bmatrix} 3 & 0 \\ 8 & -1 \end{bmatrix}$, $\vec{u} = \begin{bmatrix} 1 \\ 2 \end{bmatrix}$ *and* $\vec{v} = \begin{bmatrix} 1 \\ -1 \end{bmatrix}$. *Are* $\vec{u}$ *and* $\vec{v}$ *eigenvectors of A ?*

Solution : $\qquad A\vec{u} = \begin{bmatrix} 3 & 0 \\ 8 & -1 \end{bmatrix} \begin{bmatrix} 1 \\ 2 \end{bmatrix} = \begin{bmatrix} 3 \\ 6 \end{bmatrix} = 3 \begin{bmatrix} 1 \\ 2 \end{bmatrix}$

And $\qquad A\vec{v} = \begin{bmatrix} 3 & 0 \\ 8 & -1 \end{bmatrix} \begin{bmatrix} 1 \\ -1 \end{bmatrix} = \begin{bmatrix} 3 \\ 6 \end{bmatrix} \neq \lambda \begin{bmatrix} 1 \\ -1 \end{bmatrix}$

Thus, $\vec{u}$ is eigenvector corresponding to an eigenvalue 3, but $\vec{v}$ is not an eigenvector of A, as $A\vec{v}$ is not scalar multiple of $\vec{v}$.

Example 2.3 : *Is* $\lambda = 2$ *an eigenvalue of* $A = \begin{bmatrix} 3 & 2 \\ 3 & 8 \end{bmatrix}$? *Why or why not ? If your answer is 'yes', find the corresponding eigenvectors.*

Solution : The scalar 2 is an eigenvalue of A if and only if the equation

$$A\vec{x} = 2\vec{x} \qquad\qquad \dots \text{(i)}$$

has a non-trivial solution where, $\vec{x} = \begin{bmatrix} x_1 \\ x_2 \end{bmatrix}$. But the equation (i) is

equivalent to $\quad A\vec{x} - 2I\vec{X} = \vec{0}$

or $\qquad (A - 2I)\vec{x} = \vec{0} \qquad\qquad \dots \text{(ii)}$

So we have to solve the homogeneous system (ii), we form the matrix

$$A - 2I = \begin{bmatrix} 3 & 2 \\ 3 & 8 \end{bmatrix} - \begin{bmatrix} 2 & 0 \\ 0 & 2 \end{bmatrix} = \begin{bmatrix} 1 & 2 \\ 3 & 6 \end{bmatrix}$$

We see the columns of A $-$ 2I are linearly dependent, so (ii) has non-trivial solution. Thus, 2 is an eigenvalue of A.

Now to find the corresponding eigenvectors, we use row operations

$$\begin{bmatrix} 1 & 2 & 0 \\ 3 & 6 & 0 \end{bmatrix} \xrightarrow{R_2 + (-3)\,R_1} \begin{bmatrix} 1 & 2 & 0 \\ 0 & 0 & 0 \end{bmatrix}$$

The solution is given by $x_1 + 2x_2 = 0$ or $x_1 = -2x_2$, x_2 is free variable. Thus, the general solution is

$$\begin{bmatrix} x_1 \\ x_2 \end{bmatrix} = \begin{bmatrix} -2x_2 \\ x_2 \end{bmatrix} = x_2 \begin{bmatrix} -2 \\ 1 \end{bmatrix}$$

This shows that each vector $x_2 \begin{bmatrix} -2 \\ 1 \end{bmatrix}$ with $x_2 \neq 0$ is an eigenvector corresponding to $\lambda = 2$.

Example 2.4 : *For the matrix A in example (2.3), show that 9 is eigenvalue of A and further find the corresponding eigenvectors of $\lambda = 9$.*

Solution : The scalar $\lambda = 9$ is eigenvalue of A if and only if the equation

$$A\vec{x} = 9\vec{x} \qquad \text{... (i)}$$

has a non-trivial solution. But the equation is equivalent to

$$(A - 9I)\,\vec{x} = \vec{0} \qquad \text{... (ii)}$$

We form the matrix :

$$A - 9I = \begin{bmatrix} 3 & 2 \\ 3 & 8 \end{bmatrix} - \begin{bmatrix} 9 & 0 \\ 0 & 9 \end{bmatrix} = \begin{bmatrix} -6 & 2 \\ 3 & -1 \end{bmatrix}$$

As the columns of $A - 9I$ are linearly dependent, the equation (ii) has non-trivial solution. Thus, $\lambda = 9$ is an eigenvalue of A.

Now to find the corresponding eigenvectors, we use row operations

$$\begin{bmatrix} -6 & 2 & 0 \\ 3 & -1 & 0 \end{bmatrix} \xrightarrow{R_2 + \left(-\frac{1}{2}\right) R_1} \begin{bmatrix} -6 & 2 & 0 \\ 0 & 0 & 0 \end{bmatrix}$$

The solution is then given by $-6x_1 + x_2 = 0$ or $x_1 = \dfrac{1}{6} x_2$, x_2 is free variable, so the general solution is given by

$$\begin{bmatrix} x_1 \\ x_2 \end{bmatrix} = \begin{bmatrix} \frac{1}{6} x_2 \\ x_2 \end{bmatrix} = x_2 \begin{bmatrix} \frac{1}{6} \\ 1 \end{bmatrix}$$

This shows that each vector $x_2 \begin{bmatrix} \frac{1}{6} \\ 1 \end{bmatrix}$ is eigenvector corresponding to $\lambda = 9$ of A, with $x_2 \neq 0$.

Note :

1. The row reduction is used to find eigenvectors in the above examples, but it cannot be used to find eigenvalues.

2. From definition of eigenvalue of a matrix, if A is n × n matrix, then a scalar λ is an eigenvalue of A if and only if the equation $A\vec{x} = \lambda\vec{x}$ has a non-trivial solution. This is equivalent : λ is eigenvalue of A if and only if

   $$(A - \lambda I)\,\vec{x} = 0 \quad (*)$$

   has a non-trivial solution. We know that the set of all solutions of (*) is a null space of $(A - \lambda I)$, and is a subspace of $\mathbb{R}^n$ it is called the **eigenspace** of A corresponding to λ. Thus, eigenspace consists of a zero vector and all the eigenvectors corresponding to λ.

3. In example (2.3), eigenspace of A corresponding to $\lambda = 2$ is the set of all vectors which are scalar multiples of the vector $\begin{bmatrix} -2 \\ 1 \end{bmatrix}$, which is the line through $(-2, 1)$ and the origin.

   In example (4) the eigenspace corresponding to $\lambda = 9$ is the line through $\left(\dfrac{1}{6}, 1\right)$ and the origin. See the Fig. 2.2.

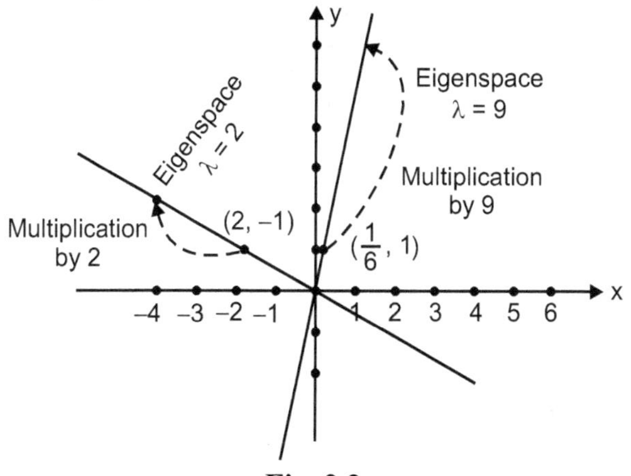

Fig. 2.2

Example 2.5 : *Let A = $\begin{bmatrix} 2 & 1 & 1 \\ 2 & 3 & 2 \\ 3 & 3 & 4 \end{bmatrix}$.*

Show that on eigenvalue of A is 1. Find the basis for the corresponding eigenspace.

Solution : Form the matrix $A - 1I$.

$$\Rightarrow \quad A - 1I = \begin{bmatrix} 2 & 1 & 1 \\ 2 & 3 & 2 \\ 3 & 3 & 4 \end{bmatrix} - 1 \begin{bmatrix} 1 & 0 & 0 \\ 0 & 1 & 0 \\ 0 & 0 & 1 \end{bmatrix} = \begin{bmatrix} 1 & 1 & 1 \\ 2 & 2 & 2 \\ 3 & 3 & 3 \end{bmatrix}$$

Now we row reduce the augmented matrix for $(A - 1I)\,\vec{x} = \vec{0}$.

$$\begin{bmatrix} 1 & 1 & 1 & 0 \\ 2 & 2 & 2 & 0 \\ 3 & 3 & 3 & 0 \end{bmatrix} \begin{array}{c} R_2 + (-2)\,R_1 \\ \sim \\ R_3 + (-3)\,R_1 \end{array} \begin{bmatrix} 1 & 1 & 1 & 0 \\ 0 & 0 & 0 & 0 \\ 0 & 0 & 0 & 0 \end{bmatrix}$$

At this stage, it is clear that $\lambda = 1$ is an eigenvalue of A, since the equation $(A - 1I)\,\vec{x} = \vec{0}$ has free variables x_2, x_3. The general solution is

$$\begin{bmatrix} x_1 \\ x_2 \\ x_3 \end{bmatrix} = \begin{bmatrix} -x_2 - x_3 \\ x_2 \\ x_3 \end{bmatrix} = x_2 \begin{bmatrix} -1 \\ 1 \\ 0 \end{bmatrix} + x_3 \begin{bmatrix} -1 \\ 0 \\ 1 \end{bmatrix}$$

Thus, the eigenspace corresponding to $\lambda = 1$ of A is clearly a two-dimensional subspace of $\mathbb{R}^3$, a plane passing through $(-1, 1, 0)$, $(-1, 0, 1)$ and the origin. A basis of this eigenspace is

$$\left\{ \begin{bmatrix} -1 \\ 1 \\ 0 \end{bmatrix}, \begin{bmatrix} -1 \\ 0 \\ 1 \end{bmatrix} \right\}$$

We now see a special case of matrices for which it is easy to find the eigenvalues, namely triangular matrices.

Theorem 1 The eigenvalues of a triangular matrix are the entries on its main diagonal.

Proof : For convenience, let us consider an upper triangular matrix of 3×3 size that is

$$A = \begin{bmatrix} a_{11} & a_{12} & a_{13} \\ 0 & a_{22} & a_{23} \\ 0 & 0 & a_{33} \end{bmatrix}$$

Then $A - \lambda I$ has the form

$$A - \lambda I = \begin{bmatrix} a_{11} & a_{12} & a_{13} \\ 0 & a_{22} & a_{23} \\ 0 & 0 & a_{33} \end{bmatrix} - \lambda \begin{bmatrix} 1 & 0 & 0 \\ 0 & 1 & 0 \\ 0 & 0 & 1 \end{bmatrix}$$

$$= \begin{bmatrix} a_{11} - \lambda & a_{12} & a_{13} \\ 0 & a_{22} - \lambda & a_{23} \\ 0 & 0 & a_{33} - \lambda \end{bmatrix}$$

Thus, λ is an eigenvalue of A if and only if the equation $(A - \lambda I)\,\vec{x} = \vec{0}$ has a non-trivial solution, that is, if and only if the equation has a free variable. Because, the last matrix is echelon form, it is easy to see that $(A - \lambda I)\,\vec{x} = \vec{0}$ has a free variable if and only if at least one of the entries on the main diagonal of $A - \lambda I$ is zero. This happens if and only if λ equals to one of the entries a_{11}, a_{22}, a_{33} in A. Thus λ is an entry on the main diagonal of A. Similarly, we can prove the case if A is lower triangular matrix. For example, let

$$A = \begin{bmatrix} 1 & 2 & 3 \\ 0 & 5 & -1 \\ 0 & 0 & 4 \end{bmatrix} \text{ and } B = \begin{bmatrix} 2 & 0 & 0 \\ -1 & 4 & 0 \\ 2 & 3 & 0 \end{bmatrix},$$

then 1, 5 and 4 are eigenvalues of A, and 2, 4, 0 are eigenvalues of B.

Note : The scalar $\lambda = 0$ - zero is an eigenvalue of a matrix if and only if the equation

$$A\vec{x} = 0\vec{x} \qquad\qquad\qquad \dots (*)$$

has a non-trivial solution. But the equation $(*)$ is equivalent to $A\vec{x} = \vec{0}$, which has a non-trivial solution if and only if A not invertible matrix. Thus, **the scalar zero $(\lambda = 0)$ is an eigenvalue of a matrix A if and only if A is not invertible (i.e. A is singular).**

The following theorem characterizes the eigenvectors corresponding to the distinct eigenvalues of a matrix.   ■

$\boxed{\textbf{Theorem 2}}$ Let A be an $n \times n$ matrix. If $\vec{u}_1, \dots, \vec{u}_k$ are eigenvectors corresponding to the distinct eigenvalues $\lambda_1, \lambda_2, \dots, \lambda_k$ of the matrix A, then the set $\{\vec{u}_1, \dots, \vec{u}_k\}$ is linearly independent.

Proof : Since $\vec{u}_1$ is an eigenvector $\vec{u}_1 \neq \vec{0}$, so if $\{\vec{u}_1, \dots, \vec{u}_k\}$ is linearly dependent there exists a vector $\vec{u}_{l+1}$, where l is the least suffix such that $\vec{u}_{l+1}$ is a linear combination of the preceding (linearly independent) vectors. Thus, there exist scalar $\alpha_1, \alpha_2, \dots, \alpha_l$ such that

$$\alpha_1 \vec{u}_1 + \alpha_2 \vec{u}_2 + \dots + \alpha_l \vec{u}_l = \vec{u}_{l+1} \qquad\qquad \dots (1)$$

Multiplying both sides of (1) by A, and using the fact that $A\vec{u}_i = \lambda_i \vec{u}_i$ for each i, we obtain

$$\alpha_1 A\vec{u}_1 + \alpha_2 A\vec{u}_2 + \dots + \alpha_l A\vec{u}_l = A\vec{u}_{l+1}$$

$$\text{or} \quad \alpha_1 \lambda_1 \vec{u}_1 + \alpha_2 \lambda_2 \vec{u}_2 + \dots + \alpha_l \lambda_l \vec{u}_l = \lambda_{l+1}\, u_{l+1} \qquad\qquad \dots (2)$$

Multiplying both sides of (1) by λ_{l+1} and subtracting the result from (2), we get

$$\alpha_1(\lambda_1 - \lambda_{l+1})\,\vec{u}_1 + \dots + \alpha_l(\lambda_l - \lambda_{l+1})\,\vec{u}_l = \vec{0} \qquad \dots (3)$$

Since, the vectors $\vec{u}_1, \dots, \vec{u}_l$ are linearly independent, the weights in (3) are all zero; that is, $\alpha_i(\lambda_i - \lambda_{l+1}) = 0$ for $i = 1, 2, \dots, l$. Since, $\lambda_1, \lambda_2, \dots, \lambda_k$ are all distinct, so $(\lambda_i - \lambda_{l+1}) \neq 0$ for $i = 1$ to l, hence $\alpha_i = 0$, for each $i = 1$ to l. But then (2) implies that $\vec{u}_{l+1} = \vec{0}$ which is impossible. Hence, $\{\vec{u}_1, \dots, \vec{u}_k\}$ must be linearly independent. $\blacksquare$

Example 2.6 : *Is* $\vec{u} = \begin{bmatrix} 4 \\ -3 \\ 1 \end{bmatrix}$ *an eigenvector of the matrix*

$A = \begin{bmatrix} 3 & 7 & 9 \\ -4 & -5 & 1 \\ 2 & 4 & 4 \end{bmatrix}$ *? If so, find the eigenvalue.*

Solution : $\vec{u}$ is eigenvector of A if and only, if $A\vec{u}$ is a scalar multiple of $\vec{u}$, so we find $A\vec{u}$.

$$A\vec{u} = \begin{bmatrix} 3 & 7 & 9 \\ -4 & -5 & 1 \\ 2 & 4 & 4 \end{bmatrix} \begin{bmatrix} 4 \\ -3 \\ 1 \end{bmatrix} = \begin{bmatrix} 0 \\ 0 \\ 0 \end{bmatrix} \neq \lambda \begin{bmatrix} 4 \\ -3 \\ 1 \end{bmatrix}$$

For any scalar λ, hence $\vec{u}$ cannot be an eigenvector of A.

Example 2.7 : *If λ is an eigenvalue of an invertible matrix A, then show that λ^{-1} is an eigenvalue of A^{-1}.*

Solution : Since, λ is an eigenvalue of A, there is a non-zero vector $\vec{x}$ such that

$$A\vec{x} = \lambda\vec{x} \qquad \dots (i)$$

Multiplying both sides by A^{-1}, we get

$$A^{-1}(A\vec{x}) = A^{-1}(\lambda\vec{x})$$

$$(A^{-1}A)\,\vec{x} = \lambda(A^{-1}\vec{x})$$

$$I\vec{x} = \lambda(A^{-1}\vec{x})$$

$$\therefore \qquad A^{-1}\vec{x} = \frac{1}{\lambda}\vec{x} = \lambda^{-1}\vec{x}$$

Thus, λ^{-1} is an eigenvalue of A^{-1}.

2.3 The Characteristic Equation

In the last section, we saw that given a scalar whether it is an eigenvalue of the matrix, but how to find an eigenvalue of a matrix is not answered. We answer this question in this section, with simple example of 2×2 matrix which will lead to the general case.

The main part that we will use here is the properties of the determinants that we studied in semester - 1.

Example 2.8 : *Find the eigenvalues of* $A = \begin{bmatrix} 1 & 1 \\ 1 & 1 \end{bmatrix}$.

Solution : We need to find scalars λ such that the matrix equation $A\vec{x} = \lambda \vec{x}$ or equivalently $(A - \lambda I)\, \vec{x} = \vec{0}$ has a non-trivial solution. By the Invertible Matrix theorem (we studied in first semester), this problem is equivalent to find all λ such that the matrix $A - \lambda I$ is **not invertible**, where in the present case

$$A - \lambda I = \begin{bmatrix} 1 & 1 \\ 1 & 1 \end{bmatrix} - \lambda \begin{bmatrix} 1 & 0 \\ 0 & 1 \end{bmatrix} = \begin{bmatrix} 1 - \lambda & 1 \\ 1 & 1 - \lambda \end{bmatrix}$$

We know that a matrix A fails to be invertible if its **determinant is zero.** Thus, the eigenvalues of A are the solutions of the equation

$$\det(A - \lambda I) = \det \begin{bmatrix} 1 - \lambda & 1 \\ 1 & 1 - \lambda \end{bmatrix} = 0$$

This implies, $(1 - \lambda)(1 - \lambda) - 1 = 0$

or $\qquad\qquad \lambda^2 - 2\lambda + 1 - 1 = 0$

or $\qquad\qquad\qquad \lambda^2 - 2\lambda = 0$

or $\qquad\qquad\qquad \lambda(\lambda - 2) = 0$

So this gives $\lambda = 0$ and $\lambda = 2$.

We state here the one important property of the determinant of a matrix.

Let A be an $n \times n$ matrix and let U be any echelon form obtained from A by row replacements and row interchanges (without scalar multiplication) and r be the number of such row interchanges. Then

$$\det A = (-1)^r \,(\text{product of pivots in U})$$

$$= (-1)\, u_{11} \cdot u_{22} \dots u_{nn}$$

where $u_{11}, u_{22}, \dots, u_{nn}$ are all pivots in U.

Note :

1. If A is **invertible,** then no u_{ii}, i = 1, 2, …, n is zero (because in this case A ~ I_n and u_{ii} have not been scaled to 1's. If A is **not invertible** then at least one u_{ii} must be zero.

Thus, $\det A = \begin{cases} (-1)^r\, u_{11} \, \ldots \, u_{nn} \, , & \text{if A is invertible} \\ 0 & , \text{if A is not invertible} \end{cases}$

2. Given a matrix A and any echelon form U obtained from A without scalar multiplications gives the same value of det A.

Example 2.9 : Find the determinant of A, where

$$A = \begin{bmatrix} 2 & 1 & 1 \\ 2 & 3 & 4 \\ -1 & -1 & -2 \end{bmatrix}$$

Solution : We row reduce A to echelon form :

$$A \xrightarrow{R_{13}} \begin{bmatrix} -1 & -1 & -2 \\ 2 & 3 & 4 \\ 2 & 1 & 1 \end{bmatrix} \xrightarrow[R_3 + 2R_1]{R_2 + 2R_1} \begin{bmatrix} -1 & -1 & -2 \\ 0 & 1 & 0 \\ 0 & -1 & -3 \end{bmatrix}$$

$$\xrightarrow{R_3 + R_2} \begin{bmatrix} -1 & -1 & -2 \\ 0 & 1 & 0 \\ 0 & 0 & -3 \end{bmatrix}$$

We have used one interchange of rows, hence

$$\det A = (-1)^1 (-1)\, 1 \cdot (-3) = -3$$

Alternatively, let us obtain echelon form U from A without interchange of row :

$$A \xrightarrow[R_3 + \frac{1}{2}R_1]{R_2 - R_1} \begin{bmatrix} 2 & 1 & 1 \\ 0 & +2 & 3 \\ 0 & \frac{-1}{2} & \frac{-3}{2} \end{bmatrix} \xrightarrow{R_3 + \frac{1}{4}R_2} \begin{bmatrix} 2 & 1 & 1 \\ 0 & 2 & 3 \\ 0 & 0 & \frac{-3}{4} \end{bmatrix}$$

Thus, $\det A = 2 \times 2 \times \left(\dfrac{-3}{4}\right) = -3$

Some More Facts About Determinant :

From the note 1, and the formula for det A, we see that **"the matrix A is invertible if and only if det A is non-zero."** Combining this fact with the Invertible Matrix Theorem, we have.

$\boxed{\text{Theorem 3}}$ Let A be an n × n matrix, the statement "A is invertible" is equivalent to the following two statements :

(i)   The number zero is not an eigenvalue of A.

(ii) The determinant of A is not zero.   ■

Recall : When A is 3×3 matrix, and if $\vec{a_1}, \vec{a_2}, \vec{a_3}$ are its columns then |det A| is the **volume** of the parallelepiped determined by $\vec{a_1}, \vec{a_2}, \vec{a_3}$. This volume is non-zero if and only if the vectors $\vec{a_1}, \vec{a_2}, \vec{a_3}$ are linearly independent and the matrix A is invertible. If $\vec{a_1}, \vec{a_2}, \vec{a_3}$ are linearly dependent, they lie in a plane.

The following theorem is proved in semester - 1, we state it only for our reference in this section and next sections.   ■

$\boxed{\text{Theorem 4}}$ **Properties of the determinants**

Let A and B be an $n \times n$ matrices :

(a)   A is invertible if and only if det A $\neq$ 0.

(b)   det (AB) = (det A) (det B)

(c)   det A^t = det A.

(d)   If A is triangular, then det A is the product of the entries on the main diagonal of A.

(e)   a row replacement operation on A does not change the determinant A; row interchange changes the sign of the determinant. α scalar multiplication also scalar multiplies the determinant by the same scalar factor.

The Characteristic Equation :

Let A be an $n \times n$ matrix, then the scalar equation det $(A - \lambda I) = 0$ is called the **characteristic equation** of A.

Remark : (1) A scalar λ is an eigenvalue of an $n \times n$ matrix A if and only if λ satisfies the characteristic equation det $(A - \lambda I) = 0$.

(2)   If A is an $n \times n$ matrix, then it can be seen that det $(A - \lambda I)$, is a **polynomial in λ of degree n;** called the **characteristic polynomial of A.** Each root of the characteristic polynomial of A is an eigenvalue of A.

(3)   A root λ of the characteristic polynomial of A is called of **multiplicity** r if $(x - \lambda)$ occurs r times as a factor of the characteristic polynomial. In general, the **algebraic multiplicity** of an eigenvalue λ is its multiplicity as a root of the characteristic polynomial.   ■

Example 2.11 : *Find the characteristic polynomial, the eigenvalues and their multiplicity for the following matrices :*

1. $A = \begin{bmatrix} 5 & -3 \\ -4 & 3 \end{bmatrix}$ 2. $A = \begin{bmatrix} 1 & 0 & -1 \\ 2 & 3 & -1 \\ 0 & 6 & 0 \end{bmatrix}$

3.
$$\begin{bmatrix} 4 & -7 & 0 & 2 \\ 0 & 3 & -4 & 6 \\ 0 & 0 & 3 & -8 \\ 0 & 0 & 0 & 1 \end{bmatrix}$$

Solution : (1)
$$A - \lambda I = \begin{bmatrix} 5 & -3 \\ -4 & 3 \end{bmatrix} - \lambda \begin{bmatrix} 1 & 0 \\ 0 & 1 \end{bmatrix}$$

$$= \begin{bmatrix} 5 - \lambda & -3 \\ -4 & 3 - \lambda \end{bmatrix}$$

$$\therefore \quad \det(A - \lambda I) = \begin{bmatrix} 5 - \lambda & -3 \\ -4 & 3 - \lambda \end{bmatrix}$$

$$= (5 - \lambda)(3 - \lambda) - 12$$

$$= 15 - 8\lambda + \lambda^2 - 12$$

$$= \lambda^2 - 8\lambda + 3$$

So the characteristic polynomial of A is $\lambda^2 - 8\lambda + 3$.

Observe that $\lambda^2 - 8\lambda + 3 = (\lambda - (4 + \sqrt{13})] [\lambda - (4 - \sqrt{13})]$

The eigenvalues are $4 + \sqrt{13}$ and $4 - \sqrt{13}$ each of multiplicity, one.

(2)
$$A - \lambda I = \begin{bmatrix} 1 & 0 & -1 \\ 2 & 3 & -1 \\ 0 & 6 & 0 \end{bmatrix} - \lambda \begin{bmatrix} 1 & 0 & 0 \\ 0 & 1 & 0 \\ 0 & 0 & 1 \end{bmatrix}$$

$$= \begin{bmatrix} 1 - \lambda & 0 & -1 \\ 2 & 3 - \lambda & -1 \\ 0 & 6 & -\lambda \end{bmatrix}$$

$$\therefore \quad \det(A - \lambda I) = \det \begin{bmatrix} 1 - \lambda & 0 & -1 \\ 2 & 3 - \lambda & -1 \\ 0 & 6 & -\lambda \end{bmatrix}$$

Using cofactor expansion along first row, we obtain

$$\det(A - \lambda I) = (1 - \lambda) \det \begin{bmatrix} 3 - \lambda & -1 \\ 6 & -\lambda \end{bmatrix} + (-1) \det \begin{bmatrix} 2 & 3 - \lambda \\ 0 & 6 \end{bmatrix}$$

$$(1 - \lambda)[(3 - \lambda)(-\lambda) + 6] + (-1)[12 - 0 \times (3 - \lambda)]$$

$$= (1 - \lambda)(-3\lambda + \lambda^2 + 6) - 12$$

$$= -\lambda^3 + 4\lambda^2 - 9\lambda + 6$$

The characteristic polynomial of A is $-\lambda^3 + 4\lambda^2 - 9\lambda + 6$.

Observe that, $-\lambda^3 + 4\lambda - 9\lambda + 6 = -(\lambda - 1)(\lambda^2 - 3\lambda + 6)$

$\lambda = 1$ is the only real root, remaining two roots are imaginary.

$\lambda = 1$ is the only one real eigenvalue of A, rest two are imaginary (complex) eigenvalues.

(3)   In this case,

$$A - \lambda I = \begin{bmatrix} 4 - \lambda & -7 & 0 & 2 \\ 0 & 3 - \lambda & -4 & 6 \\ 0 & 0 & 3 - \lambda & -8 \\ 0 & 0 & 0 & 1 - \lambda \end{bmatrix}$$

Therefore, the characteristic polynomial is

$$\det (A - \lambda I) = (4 - \lambda)(3 - \lambda)(3 - \lambda)(1 - \lambda)$$
$$= (4 - \lambda)(3 - \lambda)^2 (1 - \lambda)$$

4, 3, 1 are the eigenvalues of A, 4 and 1 are of multiplicity one each and 3 is of multiplicity 2.

Similar Matrices (Similarity)

Definition : Let A and B be two $n \times n$ matrices, then we say that **A is similar to B** if there is an invertible matrix P such that $P^{-1}AP = B$.

Note : If A is similar to B, then $P^{-1}AP = B$, which is equivalent to $A = PBP^{-1}$. If we write $P^{-1} = Q$, then $Q^{-1} = P$. So that $A = Q^{-1}BQ$, which shows that **B is similar to A.**

Therefore, we say that A and B **are similar.**

(2)   Changing A to $P^{-1}AP$ is called a **similarity transformation.**

Theorem 5 If $n \times n$ matrices A and B are similar, then they have the same characteristic polynomial and hence the same eigenvalues (with the same multiplicity).

Proof : Since, A and B are similar matrices, we have an invertible matrix P such that $B = P^{-1}AP$, then

$$B - \lambda I = P^{-1}AP - \lambda I$$
$$= P^{-1}AP - \lambda P^{-1}P \qquad \because P^{-1}P = I$$
$$= P^{-1}AP - P^{-1}(\lambda I) P$$
$$= P^{-1}(A - \lambda I) P.$$

Using multiplicative property of the determinants, we have

$$\det (B - \lambda I) = \det [P^{-1}(A - \lambda I)P]$$
$$= \det P^{-1} \cdot \det (A - \lambda I) \cdot \det P$$
$$= \det (A - \lambda I)$$

Since, $\det P^{-1} \cdot \det P = 1$

Thus, $\det (B - \lambda I) = \det (A - \lambda I)$

∎

Example 2.12 : *Prove that for a square matrix A, A and A^t have the same characteristic polynomial.*

Solution : We know that $\det A^t = \det A$.

Using this fact, we have

$$\det (A^t - \lambda I) = \det (A^t - \lambda I)^t$$

But $\qquad (A^t - \lambda I)^t = (A^t)^t - (\lambda I)^t = A - \lambda I$

Hence, $\qquad \det (A^t - \lambda I) = \det (A - \lambda I)$

Example 2.13 : *Let A be $n \times n$ matrix, and suppose A has n real eigenvalues, $\lambda_1, \lambda_2, ..., \lambda_n$, repeated according to the multiplicities, so that $\det (A - \lambda I) = (\lambda_1 - \lambda) ... (\lambda_n - \lambda)$. Show that $\det A$ is the product of the n eigenvalues of A.*

Solution : It is clear that, if we see the right side of

$$\det (A - \lambda I) = (\lambda_1 - \lambda)(\lambda_2 - \lambda) ... (\lambda_n - \lambda),$$

we will get a polynomial in λ of degree n and with the constant term to be $\lambda_1 \lambda_2 ... \lambda_n$. Thus, $\det (A - \lambda I) = \lambda_1 \lambda_2 ... \lambda_n + (*) \lambda + ... + (*) \lambda^n ...$ (1)

where $(*)$ denoted the coefficients of $\lambda, \lambda^2, ..., \lambda^n$. Now in equation (1) if we put $\lambda = 0$ on both sides we get

$$\det (A - 0) = \lambda_1 \lambda_2 ... \lambda_n + 0 + ... + 0$$

That is $\det A = \lambda_1 \lambda_2 ... \lambda_n$.

2.4 Diagonalization

In this section, we see that the applications of eigenvalues and eigenvectors in particular, given a square matrix A, we can find A^k, for large values of integer $k \geq 0$ using these two concepts. For this we need to define a diagonal matrix.

Diagonal matrix : A square matrix A is said to be diagonal if all its non-diagonal entries are zeros.

For example, $\begin{bmatrix} 5 & 0 \\ 0 & -1 \end{bmatrix}, \begin{bmatrix} 1 & 0 & 0 \\ 0 & 3 & 0 \\ 0 & 0 & 7 \end{bmatrix}, \begin{bmatrix} -2 & 0 & 0 & 0 \\ 0 & 4 & 0 & 0 \\ 0 & 0 & 0 & 0 \\ 0 & 0 & 0 & 12 \end{bmatrix}$ are diagonal

matrices 2×2, 3×3 and 4×4 size.

Note that if a square matrix A is diagonal it is easy to find A^k, for any $k \geq 0$.

Example 2.14 : *If $A = \begin{bmatrix} 5 & 0 \\ 0 & -1 \end{bmatrix}$, find A^k, $k \geq 0$.*

Solution : We have by matrix multiplication

$$A^2 = AA = \begin{bmatrix} 5 & 0 \\ 0 & -1 \end{bmatrix}\begin{bmatrix} 5 & 0 \\ 0 & -1 \end{bmatrix}$$

$$= \begin{bmatrix} 5 \times 5 + 0 \times 0 & 5 \times 0 + 0\,(-1) \\ 0 \times 5 + (-1) \times 0 & 0 \times 0 + (-1)\,(-1) \end{bmatrix}$$

$$= \begin{bmatrix} 5^2 & 0 \\ 0 & (-1)^2 \end{bmatrix}$$

$$A^3 = AA^2 = \begin{bmatrix} 5 & 0 \\ 0 & -1 \end{bmatrix}\begin{bmatrix} 5^2 & 0 \\ 0 & (-1)^2 \end{bmatrix}$$

$$= \begin{bmatrix} 5^3 & 0 \\ 0 & (-1)^3 \end{bmatrix}$$

So in general, $\qquad A^k = \begin{bmatrix} 5^k & 0 \\ 0 & (-1)^k \end{bmatrix}$

Definition : A square matrix A is said to be **diagonalizable** if A is similar to a diagonal matrix. That is, if $A = PDP^{-1}$ for some invertible matrix P and for some diagonal matrix D.

The following gives the criterion for a square matrix to be diagonalizable.

Theorem 6 (The Diagonal Matrix Theorem)

An $n \times n$ matrix A is diagonalizable if and only if A has n linearly independent eigenvectors.

In other words, $A = PDP^{-1}$, with D diagonal matrix, if and only if the columns of P are n-linearly independent eigenvectors of A. In this case, the diagonal entries of D are the corresponding eigenvalues respectively.

Proof : Let $\vec{u}_1, \vec{u}_2, \ldots, \vec{u}_n$ be any n-vectors in $\mathbb{R}^n$ and $\lambda_1, \lambda_2, \ldots, \lambda_n$ be n scalars. Let P be $n \times n$ matrix with columns $\vec{u}_1, \vec{u}_2, \ldots, \vec{u}_n$ and D be a diagonal matrix with diagonal entries $\lambda_1, \lambda_2, \ldots, \lambda_n$. Thus,

$$P = [\vec{u}_1, \vec{u}_2, \ldots, \vec{u}_n] \text{ and}$$

$$D = \begin{bmatrix} \lambda_1 & 0 & - & - & - & - & 0 \\ 0 & \lambda_2 & - & - & - & - & 0 \\ & & & & & & \\ 0 & 0 & - & - & - & - & \lambda_n \end{bmatrix}$$

Then we have

$$AP = A[\vec{u}_1 \; \vec{u}_2 \ldots \vec{u}_n] = [A\vec{u}_1 \; A\vec{u}_2 \ldots A\vec{u}_n] \qquad \ldots (1)$$

while $\qquad PD = [\vec{u}_1 \; \vec{u}_2 \ldots \vec{u}_n] \, D$

$$\therefore \qquad PD = [\lambda_1\vec{u}_1 \; \lambda_2\vec{u}_2 \ldots \lambda_n\vec{u}_n] \qquad \ldots (2)$$

Now, let us suppose that A is diagonalizable, so that $A = PDP^{-1}$, then right multiplying both sides by P, we get

$$AP = PD$$

In this case, (1) and (2) imply that

$$[A\vec{u}_1 \; A\vec{u}_2 \ldots A\vec{u}_n] = [\lambda_1\vec{u}_1 \; \lambda_2\vec{u}_2 \ldots \lambda_n\vec{u}_n] \qquad \ldots (3)$$

By equality of matrices (3) gives,

$$A\vec{u}_1 = \lambda_1\vec{u}_1, \; A\vec{u}_2 = \lambda_2\vec{u}_2, \; \ldots \; A\vec{u}_n = \lambda_n\vec{u}_n \qquad \ldots (4)$$

Since, P is invertible matrix its columns $\vec{u}_1, \vec{u}_2, \ldots, \vec{u}_n$ must be linearly independent. Showing that $\vec{u}_1, \vec{u}_2, \ldots, \vec{u}_n$ are eigenvectors corresponding to the eigenvalue $\lambda_1, \lambda_2, \ldots, \lambda_n$.

Conversely, suppose given n eigenvectors $\vec{u}_1, \vec{u}_2, \ldots, \vec{u}_n$, let P be $n \times n$ matrix whose columns are $\vec{u}_1, \vec{u}_2, \ldots, \vec{u}_n$ and let D be the diagonal entries $\lambda_1, \lambda_2, \ldots, \lambda_n$ which eigenvalues corresponding to the eigenvectors $\vec{u}_1, \vec{u}_2, \ldots, \vec{u}_n$. Then from equations (1), (2) and (3) AP = PD. This is true without any condition on the eigenvectors. But by the Invertible Matrix Theorem, if the eigenvectors $\vec{u}_1, \ldots, \vec{u}_n$ are linearly independent, then the matrix P is invertible and AP = PD implies that $A = PDP^{-1}$. ∎

Example 2.15 : *Diagonalize the matrix* $A = \begin{bmatrix} -1 & 4 & -2 \\ -3 & 4 & 0 \\ -3 & 1 & 3 \end{bmatrix}$. *That is,*

(i) find and invertible matrix P, and

(ii) diagonal matrix D such that $A = PDP^{-1}$.

Solution : We solve this example in 4 steps :

Step 1 : Finding the eigenvalues of A :

First we find the characteristic polynomial det $(A - \lambda I)$ of A.

Form the matrix $A - \lambda I$.

$$A - \lambda I = \begin{bmatrix} -1 & 4 & -2 \\ -3 & 4 & 0 \\ -3 & 1 & 3 \end{bmatrix} - \lambda \begin{bmatrix} 1 & 0 & 0 \\ 0 & 1 & 0 \\ 0 & 0 & 1 \end{bmatrix}$$

$$= \begin{bmatrix} -1 - \lambda & 4 & -2 \\ -3 & 4 - \lambda & 0 \\ -3 & 1 & 3 - \lambda \end{bmatrix} \qquad \ldots \text{(I)}$$

To find det $(A - \lambda I)$, we use expand the determinant along first row, so we have,

$$\det (A - \lambda I) = (-1 - \lambda) \det \begin{bmatrix} 4 - \lambda & 0 \\ 1 & 3 - 1 \end{bmatrix}$$

$$- 4 \det \begin{bmatrix} -3 & 0 \\ -3 & 3 - \lambda \end{bmatrix} - 2 \det \begin{bmatrix} -3 & 4 - \lambda \\ -3 & 1 \end{bmatrix}$$

$$= (-1 - \lambda)(4 - \lambda)(3 - \lambda) - 4\lambda(-3)(3 - \lambda)$$

$$= -(\lambda + 1)(\lambda - 4)(\lambda - 3) + 12(3 - \lambda) - 18 + 6\lambda$$

$$= -(\lambda^3 - 6\lambda^2 + 5\lambda + 12) + 36 - 12\lambda - 18 + 6\lambda$$

$$= -\lambda^3 + 6\lambda^2 - 5\lambda - 12 + 18 - 6\lambda$$

$$= -\lambda^3 + 6\lambda^2 - 11\lambda + 6$$

Thus, the characteristic polynomial of A is

$$\det (A - \lambda I) = -\lambda^3 + 6\lambda^2 + 11\lambda + 6$$

To find the eigenvalues, we equate det $(A - \lambda I) = 0$ and factorise the left hand side that is

$$-\lambda^3 + 6\lambda^2 - 11\lambda + 6 = 0$$

or $\quad \lambda^3 - 6\lambda^2 + 11\lambda - 6 = 0$

Using the methods of factorizing, we obtain :

$$\lambda^3 - 1 - 6\lambda^2 + 11\lambda - 5 = (\lambda - 1)[\lambda^2 + \lambda + 1 - 6\lambda + 11\lambda - 5] = 0$$

$$(\lambda - 1)(\lambda - 2)(\lambda - 3) = 0$$

Thus, the eigenvalues of A are 1, 2, 3.

Step 2 : To find eigenvectors corresponding to the eigenvalues $\lambda = 1, 2, 3$. For this we have to solve the system

$$(A - \lambda I)\, \vec{x} = \vec{0}, \ \vec{x} \in \mathbb{R}^3 \ \text{ for each } \lambda = 1, 2, 3.$$

(i) For $\lambda = 1$, we have from (I) above

$$\begin{bmatrix} -1 - 1 & 4 & -2 \\ -3 & 4 - 1 & 0 \\ -3 & 1 & 3 - 1 \end{bmatrix} \begin{bmatrix} x_1 \\ x_2 \\ x_3 \end{bmatrix} = \begin{bmatrix} 0 \\ 0 \\ 0 \end{bmatrix}$$

or
$$\begin{bmatrix} -2 & 4 & -2 \\ -3 & 3 & 0 \\ -3 & 1 & 2 \end{bmatrix} \begin{bmatrix} x_1 \\ x_2 \\ x_3 \end{bmatrix} = \begin{bmatrix} 0 \\ 0 \\ 0 \end{bmatrix} \qquad \ldots \text{(II)}$$

We reduce row of the coefficient matrix

$$\begin{bmatrix} -2 & 4 & -2 \\ -3 & 3 & 0 \\ -3 & 1 & 2 \end{bmatrix} \xrightarrow{\left(-\frac{1}{2}R_2\right)} \begin{bmatrix} 1 & 2 & 1 \\ 3 & 3 & 0 \\ -3 & 1 & 2 \end{bmatrix} \begin{matrix} R_2 + 3R_1 \\ \\ R_3 + 3R_1 \end{matrix}$$

$$\sim \begin{bmatrix} 1 & -2 & 1 \\ 0 & -3 & 3 \\ 0 & -5 & 5 \end{bmatrix} \begin{matrix} -\frac{1}{3}R_2 \\ \\ R_3 + 5R_2 \end{matrix} \begin{bmatrix} 1 & -2 & 1 \\ 0 & 1 & -1 \\ 0 & 0 & 0 \end{bmatrix}$$

Thus, (II) takes the form
$$\begin{bmatrix} 1 & -2 & 1 \\ 0 & 1 & -1 \\ 0 & 0 & 0 \end{bmatrix} \begin{bmatrix} x_1 \\ x_2 \\ x_3 \end{bmatrix} = \begin{bmatrix} 0 \\ 0 \\ 0 \end{bmatrix}$$

or
$$x_1 - 2x_2 + x_3 = 0$$
$$x_2 - x_3 = 0$$

The general solution is given by,
$$\begin{bmatrix} x_1 \\ x_2 \\ x_3 \end{bmatrix} = \begin{bmatrix} x_3 \\ x_3 \\ x_3 \end{bmatrix}$$

So, in particular $\vec{u}_1 = \begin{bmatrix} 1 \\ 1 \\ 1 \end{bmatrix}$ is the eigenvector of A corresponding to $\lambda = 1$.

(ii) For $\lambda = 2$, in (I), we have,
$$\begin{bmatrix} -1-2 & 4 & -2 \\ -3 & 4-2 & 0 \\ -3 & 1 & 3-2 \end{bmatrix} \begin{bmatrix} x_1 \\ x_2 \\ x_3 \end{bmatrix} = \begin{bmatrix} 0 \\ 0 \\ 0 \end{bmatrix}$$

or
$$\begin{bmatrix} -3 & 4 & -2 \\ -3 & 2 & 0 \\ -3 & 1 & 1 \end{bmatrix} \begin{bmatrix} x_1 \\ x_2 \\ x_3 \end{bmatrix} = \begin{bmatrix} 0 \\ 0 \\ 0 \end{bmatrix} \qquad \ldots \text{(III)}$$

We row reduce the coefficient matrix in (III).

$$\begin{bmatrix} -3 & 4 & -2 \\ -3 & 2 & 0 \\ -3 & 1 & 1 \end{bmatrix} \begin{matrix} R_2 - R_1 \\ \\ R_3 - R_1 \end{matrix} \begin{bmatrix} -3 & 4 & -2 \\ 0 & -2 & 2 \\ 0 & -3 & 3 \end{bmatrix} \sim \begin{bmatrix} -3 & 4 & -2 \\ 0 & 1 & -1 \\ 0 & 0 & 0 \end{bmatrix}$$

Thus, (III) takes the form
$$\begin{bmatrix} -3 & 4 & -2 \\ 0 & 1 & -1 \\ 0 & 0 & 0 \end{bmatrix} \begin{bmatrix} x_1 \\ x_2 \\ x_3 \end{bmatrix} = \begin{bmatrix} 0 \\ 0 \\ 0 \end{bmatrix}$$

or $\qquad -3x_1 + 4x_2 - 2x_3 = 0$

$$x_2 - x_3 = 0$$

or $\qquad x_1 = +\dfrac{2}{3}x_3$

$$x_2 = x_3$$

Therefore, the general solution of (III) is given by

$$\begin{bmatrix} x_1 \\ x_2 \\ x_3 \end{bmatrix} = \begin{bmatrix} \dfrac{2}{3}x_3 \\ x_3 \\ x_3 \end{bmatrix}$$

In particular, with $x_3 = 3$, we see that $\vec{u}_2 = \begin{bmatrix} 2 \\ 3 \\ 3 \end{bmatrix}$ is the eigenvector of

A corresponding $\lambda = 2$.

(iii) For $\lambda = 3$, in (I) we get

$$\begin{bmatrix} -1-3 & 4 & -2 \\ -3 & 4-3 & 0 \\ -3 & 1 & 0 \end{bmatrix} \begin{bmatrix} x_1 \\ x_2 \\ x_3 \end{bmatrix} = \begin{bmatrix} 0 \\ 0 \\ 0 \end{bmatrix}$$

or $\qquad \begin{bmatrix} -4 & 4 & -2 \\ -3 & 1 & 0 \\ -3 & 1 & 0 \end{bmatrix} \begin{bmatrix} x_1 \\ x_2 \\ x_3 \end{bmatrix} = \begin{bmatrix} 0 \\ 0 \\ 0 \end{bmatrix}$ $\qquad\qquad\qquad$ ... (IV)

Row reduce the coefficient matrix

$$\begin{bmatrix} -4 & 4 & -2 \\ -3 & 1 & 0 \\ -3 & 1 & 0 \end{bmatrix} \xrightarrow[R_3 - R_2]{\left(-\frac{1}{4}R_1\right)} \begin{bmatrix} 1 & -1 & 1/2 \\ -3 & 1 & 0 \\ 0 & 0 & 0 \end{bmatrix} \xrightarrow{R_2 + 3R_1} \begin{bmatrix} 1 & -1 & 1/2 \\ 0 & -2 & 3/2 \\ 0 & 0 & 0 \end{bmatrix}$$

Using this in (IV) we obtain,

$$\begin{bmatrix} 1 & -1 & 1/2 \\ 0 & -2 & 3/2 \\ 0 & 0 & 0 \end{bmatrix} \begin{bmatrix} x_1 \\ x_2 \\ x_3 \end{bmatrix} = \begin{bmatrix} 0 \\ 0 \\ 0 \end{bmatrix}$$

or $\qquad x_1 - x_2 + \dfrac{1}{2}x_3 = 0$

$$-2x_2 + \dfrac{3}{2}x_3 = 0$$

or $\qquad x_1 = x_2 - \dfrac{1}{2}x_3$

$$x_2 = \dfrac{3}{4}x_3$$

or
$$x_1 = \frac{1}{4} x_3$$

$$x_2 = \frac{3}{4} x_3$$

Thus, the general solution of (IV) is

$$\begin{bmatrix} x_1 \\ x_2 \\ x_3 \end{bmatrix} = \begin{bmatrix} \frac{1}{4} x_3 \\ \frac{3}{4} x_3 \\ x_3 \end{bmatrix}$$

Thus, in particular with $x_3 = 4$, $\vec{u}_3 = \begin{bmatrix} 1 \\ 3 \\ 4 \end{bmatrix}$ is the eigenvector of A corresponding to $\lambda = 3$.

Step 3 : Construction of matrix P.

Using the eigenvectors found in (i), (ii) and (iii).

We find, $P = [\vec{u}_1 \ \vec{u}_2 \ \vec{u}_3]$, or $P = \begin{bmatrix} 1 & 2 & 1 \\ 1 & 3 & 3 \\ 1 & 3 & 4 \end{bmatrix}$

Step 4 : Construction of the matrix D from the corresponding eigenvalues, we have,

$$D = \begin{bmatrix} 1 & 0 & 0 \\ 0 & 2 & 0 \\ 0 & 0 & 3 \end{bmatrix}$$

Remark : It is good idea to check that P and D really work for this we verify $AP = PD$.

This will ensure that our calculations were all correct.

Example 2.16 : *Diagonalize the matrix* $A = \begin{bmatrix} 4 & 0 & 0 \\ 1 & 4 & 0 \\ 0 & 0 & 5 \end{bmatrix}$.

Solution : Step 1 : To find the characteristic polynomial, it is easy in this case, as the matrix A is lower triangular, so the eigenvalues are 4, 4, 5 and the characteristic polynomial

$$\det (A - \lambda I) = (4 - \lambda)^2 (5 - \lambda)$$

Step 2 : To find eigenvector of A corresponding to $\lambda = 4$, we have

$$\begin{bmatrix} 4-4 & 0 & 0 \\ 1 & 4-4 & 0 \\ 0 & 0 & 5-4 \end{bmatrix} \begin{bmatrix} x_1 \\ x_2 \\ x_3 \end{bmatrix} = \begin{bmatrix} 0 \\ 0 \\ 0 \end{bmatrix}$$

$$\text{or} \quad \begin{bmatrix} 0 & 0 & 0 \\ 1 & 0 & 0 \\ 0 & 0 & 1 \end{bmatrix} \begin{bmatrix} x_1 \\ x_2 \\ x_3 \end{bmatrix} = \begin{bmatrix} 0 \\ 0 \\ 0 \end{bmatrix}$$

The general solution is

$$\begin{bmatrix} x_1 \\ x_2 \\ x_3 \end{bmatrix} = \begin{bmatrix} 0 \\ x_2 \\ 0 \end{bmatrix} \text{ so } \begin{bmatrix} 0 \\ 1 \\ 0 \end{bmatrix} \text{ is the eigenvector corresponding to } \lambda = 4.$$

For $\lambda = 5$, we have

$$\begin{bmatrix} 4-5 & 0 & 0 \\ 1 & 4-5 & 0 \\ 0 & 0 & 5-5 \end{bmatrix} \begin{bmatrix} x_1 \\ x_2 \\ x_3 \end{bmatrix} = \begin{bmatrix} 0 \\ 0 \\ 0 \end{bmatrix}$$

$$\text{or} \quad \begin{bmatrix} -1 & 0 & 0 \\ 1 & -1 & 0 \\ 0 & 0 & 0 \end{bmatrix} \begin{bmatrix} x_1 \\ x_2 \\ x_3 \end{bmatrix} = \begin{bmatrix} 0 \\ 0 \\ 0 \end{bmatrix}$$

This gives general solution as $x_1 = 0$, $x_2 = 0$ and x_3 free variable.

So $\begin{bmatrix} 0 \\ 0 \\ 1 \end{bmatrix}$ is the eigenvector of A corresponding to $\lambda = 5$.

Step 3 : Construction of the matrix P :

Since, we have only **two** eigenvectors corresponding to $\lambda = 4, 4, 5$, we cannot have matrix P.

Hence, the matrix A is not diagonalizable.

The following theorem provides sufficient condition for a matrix to be diagonalizable. ∎

Theorem 7 An $n \times n$ matrix is diagonalizable if it has n-distinct eigenvalues.

Proof : Let $\vec{u}_1, \vec{u}_2, ..., \vec{u}_n$ be the eigenvectors corresponding to the n-distinct eigenvalues of A. then $\{\vec{u}_1, \vec{u}_2, ..., \vec{u}_n\}$ is linearly independent by theorem 2. Hence, A is diagonalizable by theorem 5. ∎

Remark :

(1) It is not necessary for an $n \times n$ matrix A to have n distinct eigenvalues in order to be diagonalizable. The following example (16) shows that even the eigenvalues of A are not distinct, still the matrix A is diagonalizable.

(2) In example (15), the eigenvalues of the matrix A are all distinct hence by theorem (3) the eigenvalues of A corresponding to the distinct eigenvalues are linearly independent, hence A is diagonalizable.

Example 2.17 : Determine whether the matrix A is diagonalizable, where

$$A = \begin{bmatrix} 7 & 4 & 16 \\ 2 & 5 & 8 \\ -2 & -2 & -5 \end{bmatrix}.$$

We have, $A - \lambda I = \begin{bmatrix} 7-\lambda & 4 & 16 \\ 2 & 5-\lambda & 8 \\ -2 & -2 & -5-\lambda \end{bmatrix}$

So, that $\det(A - \lambda I) = \det \begin{bmatrix} 7-\lambda & 4 & 16 \\ 2 & 5-\lambda & 8 \\ -2 & -2 & -5-\lambda \end{bmatrix}$

Expanding along first row :

$$\begin{aligned}
\det(A - \lambda I) &= (7-\lambda)[(5-\lambda)(-5-\lambda) + 16] \\
&\quad - 4[2(-5-\lambda) + 16] + 16[-4 + 2(5-\lambda)] \\
&= (7-\lambda)(5-\lambda)(-5-\lambda) + (7-\lambda) \times 16 \\
&\quad - 4(6 - 2\lambda) + (96 - 32\lambda) \\
&= -(\lambda - 7)(\lambda - 5)(\lambda + 5) + 112 - 16\lambda \\
&\quad - 24 + 8\lambda + 96 - 32\lambda \\
&= -(\lambda - 7)(\lambda^2 - 25) - 40\lambda + 184 \\
&= -\lambda^3 + 7\lambda^2 - 15\lambda + 9
\end{aligned}$$

Thus, the characteristic polynomial of A is

$$\det(A - \lambda I) = -\lambda^3 + 7\lambda^2 - 15\lambda + 9$$

We factorize this polynomial, we get

$$\begin{aligned}
\det(A - \lambda I) &= -[\lambda^3 - 7\lambda^2 - 15\lambda + 9] \\
&= -[\lambda^3 - 1 - (7\lambda^2 + 15\lambda - 8)] \\
&= -[(\lambda - 1)(\lambda^2 + \lambda + 1) - (\lambda - 1)(7\lambda - 8)] \\
&= -(\lambda - 1)(\lambda^2 + \lambda + 1 - 7\lambda + 8) \\
&= -(\lambda - 1)(\lambda - 3)^2
\end{aligned}$$

$\det(A - \lambda I) = 0$ gives the eigenvalues of A as $\lambda = 1, 3, 3$.

Now to find the eigenvectors corresponding to $\lambda = 1, 3, 3$.

(i)   For $\lambda = 1$, we have to solve

$$\begin{bmatrix} 7-1 & 4 & 16 \\ 2 & 5-1 & 8 \\ -2 & -2 & -5-1 \end{bmatrix} \begin{bmatrix} x_1 \\ x_2 \\ x_3 \end{bmatrix} = \begin{bmatrix} 0 \\ 0 \\ 0 \end{bmatrix}$$

or $\begin{bmatrix} 6 & 4 & 16 \\ 2 & 4 & 8 \\ -2 & -2 & -6 \end{bmatrix} \begin{bmatrix} x_1 \\ x_2 \\ x_3 \end{bmatrix} = \begin{bmatrix} 0 \\ 0 \\ 0 \end{bmatrix}$

Row reducing the coefficient matrix, we obtain,

$$\begin{bmatrix} 1 & 2 & 4 \\ 0 & 1 & 1 \\ 0 & 0 & 0 \end{bmatrix} \begin{bmatrix} x_1 \\ x_2 \\ x_3 \end{bmatrix} = \begin{bmatrix} 0 \\ 0 \\ 0 \end{bmatrix}$$

$$x_1 + 2x_2 + 4x_3 = 0 \qquad \text{and} \qquad x_2 + x_3 = 0$$

$$x_1 + 2x_3 + 4x_3 = 0 \qquad \text{and} \qquad x_2 = -x_3$$

or $x_1 = -2x_3$ and $x_2 = -x_3$

Thus, the general solution of the system is

$$\begin{bmatrix} x_1 \\ x_2 \\ x_3 \end{bmatrix} = \begin{bmatrix} -2x_3 \\ -x_3 \\ x_3 \end{bmatrix} = x_3 \begin{bmatrix} -2 \\ -1 \\ 1 \end{bmatrix}$$

Thus, the basis for the eigenspace for $\lambda = 1$ is $\vec{u} = \begin{bmatrix} -2 \\ -1 \\ 0 \end{bmatrix}$.

(ii) For $\lambda = 3$, we have

$$\begin{bmatrix} 7-3 & 4 & 16 \\ 2 & 2 & 8 \\ -2 & -2 & -5-3 \end{bmatrix} \begin{bmatrix} x_1 \\ x_2 \\ x_3 \end{bmatrix} = \begin{bmatrix} 0 \\ 0 \\ 0 \end{bmatrix}$$

or $$\begin{bmatrix} 4 & 4 & 16 \\ 2 & 2 & 8 \\ -2 & -2 & -8 \end{bmatrix} \begin{bmatrix} x_1 \\ x_2 \\ x_3 \end{bmatrix} = \begin{bmatrix} 0 \\ 0 \\ 0 \end{bmatrix}$$

Row reducing the coefficient matrix, we obtain

$$\begin{bmatrix} 1 & 1 & 4 \\ 0 & 0 & 0 \\ 0 & 0 & 0 \end{bmatrix} \begin{bmatrix} x_1 \\ x_2 \\ x_3 \end{bmatrix} = \begin{bmatrix} 0 \\ 0 \\ 0 \end{bmatrix}$$

This gives, $x_1 + x_2 + 4x_3 = 0$

or $$x_1 = -x_2 - 4x_3$$

Thus, the general solution is

$$\begin{bmatrix} x_1 \\ x_2 \\ x_3 \end{bmatrix} = \begin{bmatrix} -x_2 - 4x_3 \\ x_2 \\ x_3 \end{bmatrix} = x_2 \begin{bmatrix} -1 \\ 1 \\ 0 \end{bmatrix} + x_3 \begin{bmatrix} -4 \\ 0 \\ 1 \end{bmatrix}$$

Therefore, the basis for the eigenspace of A for $\lambda = 3$ is

$$\vec{u}_2 = \begin{bmatrix} -1 \\ 1 \\ 0 \end{bmatrix} \text{ and } \vec{u}_3 = \begin{bmatrix} -4 \\ 0 \\ 1 \end{bmatrix}.$$

The matrix P is thus formed as

$$P = \begin{bmatrix} -2 & -1 & -4 \\ -1 & 1 & 0 \\ 1 & 0 & 1 \end{bmatrix}$$

and the matrix D is

$$D = \begin{bmatrix} 1 & 0 & 0 \\ 0 & 3 & 0 \\ 0 & 0 & 3 \end{bmatrix}$$

There are three linearly independent eigenvectors for A corresponding to the eigenvalues $\lambda = 1, 3, 3$, hence **A is diagonalizable.**

The following theorem characterizes the different cases of the matrix to be diagonalizable or not.

$\boxed{\textbf{Theorem 8}}$ Let A be an $n \times n$ matrix with distinct eigenvalues $\lambda_1, \lambda_2, \ldots, \lambda_k$.

(a) For $1 \le i \le k$, the dimension of the eigenspace for λ_i is less than or equal to the multiplicity of the eigenvalue λ_i.

(b) The matrix A is diagonalizable if and only if the sum of the dimensions of the distinct eigenspaces equals n, and this happens if and only if the dimension of the eigenspace for each λ_i equals the multiplicity of λ_i.

(c) If A is diagonalizable and β_i is a basis for the eigenspace corresponding to λ_i for each i, then the total collection of the vectors in the sets $\beta_1, \beta_2, \ldots, \beta_k$ forms an eigenvector basis for $\mathbb{R}^n$.   ■

Remark :

(1) We have seen how the higher exponent of the diagonal matrix is easy to find. This fact helps to find higher power of the given matrix, provided that it is diagonalizable.

(2) Suppose A is $n \times n$ matrix and P is invertible matrix such that $A = PDP^{-1}$, where D is diagonal matrix. Then,

$$
\begin{aligned}
A^2 = A \cdot A &= (PDP^{-1})(PDP^{-1}) \\
&= (PD)(P^{-1}P)(DP^{-1}) \\
&= (PD)\,I\,(DP^{-1}) \qquad\qquad \because P^{-1}P = I_n \\
&= P(DD)\,P^{-1} \\
&= PD^2 P^{-1}
\end{aligned}
$$

Next $\qquad A^3 = A^2 A = (PD^2P^{-1})\,(PDP^{-1})$

$$= PD^3P^{-1}$$

So, in general for any integer $k \geq 1$, we can show that

$$A^k = PD^k\,P^{-1}$$

We know D^k, easy to work out as it is diagonal, so A^k is obtained by the multiplication of the three matrices on right hand side.

Example 2.18 : *Let $A = PDP^{-1}$, compute A^4, where,*

$$P = \begin{bmatrix} 2 & -3 \\ -3 & 5 \end{bmatrix}, \quad D = \begin{bmatrix} 1 & 0 \\ 0 & 1/2 \end{bmatrix}$$

Solution : From remark (2) above, we have

$$A^4 = PD^4P^{-1}, \quad D^4 = \begin{bmatrix} 1 & 0 \\ 0 & 1/16 \end{bmatrix}$$

$\therefore \qquad A^4 = \begin{bmatrix} 2 & -3 \\ -3 & 5 \end{bmatrix} \begin{bmatrix} 1 & 0 \\ 0 & 1/16 \end{bmatrix} \begin{bmatrix} 5 & 3 \\ 3 & 2 \end{bmatrix}$

$$\because P^{-1} = \begin{bmatrix} 5 & 3 \\ 3 & 2 \end{bmatrix}$$

$$= \begin{bmatrix} 2 & \dfrac{-3}{16} \\[2ex] -3 & \dfrac{5}{16} \end{bmatrix} \begin{bmatrix} 5 & 3 \\ 3 & 2 \end{bmatrix}$$

$$= \begin{bmatrix} \dfrac{161}{16} & \dfrac{90}{16} \\[2ex] \dfrac{-225}{16} & \dfrac{-133}{16} \end{bmatrix}$$

Example 2.19 : *Using the factorization $A = PDP^{-1}$ to compute A^k, for arbitrary positive integer k, where*

$$A = \begin{bmatrix} -2 & 12 \\ -1 & 5 \end{bmatrix}, \quad P = \begin{bmatrix} 3 & 4 \\ 1 & 1 \end{bmatrix}, \quad D = \begin{bmatrix} 2 & 0 \\ 0 & 1 \end{bmatrix}$$

Solution: Using remarks (2) above, we have for any integer $k \geq 1$.

$$A^k = PD^k\,P^{-1}, \quad \text{and} \quad D^k = \begin{bmatrix} 2^k & 0 \\ 0 & 1^k \end{bmatrix}$$

So that, $\qquad A^k = \begin{bmatrix} 3 & 4 \\ 1 & 1 \end{bmatrix} \begin{bmatrix} 2^k & 0 \\ 0 & 1^k \end{bmatrix} \begin{bmatrix} -1 & 4 \\ 1 & -3 \end{bmatrix}$

$$= \begin{bmatrix} 3 \times 2^k & 4 \\ 2^k & 1 \end{bmatrix} \begin{bmatrix} -1 & 4 \\ 1 & -3 \end{bmatrix}$$

$$= \begin{bmatrix} -3 \times 2^k + 4 & 3 \times 4 \times 2^k - 12 \\ -2^k + 1 & 4 \times 2^k - 3 \end{bmatrix}$$

$$\therefore \qquad A^k = \begin{bmatrix} -3 \times 2^k + 4 & 12(2^k - 1) \\ -2^k + 1 & 4 \times 2^k - 3 \end{bmatrix}$$

Example 2.20 : The matrix A is factored in the form PDP^{-1}, using the diagonalization theorem to find eigenvalues of A and basis for each eigenspace, where $A = PDP^{-1}$ is

$$\begin{bmatrix} 2 & 1 & 1 \\ 1 & 3 & 1 \\ 1 & 2 & 2 \end{bmatrix} = \begin{bmatrix} 1 & 1 & 2 \\ 1 & 0 & -1 \\ 1 & -1 & 0 \end{bmatrix} \begin{bmatrix} 5 & 0 & 0 \\ 0 & 1 & 0 \\ 0 & 0 & 1 \end{bmatrix} \begin{bmatrix} \frac{1}{4} & \frac{1}{2} & \frac{1}{4} \\ \frac{1}{4} & \frac{1}{2} & \frac{-3}{4} \\ \frac{1}{4} & \frac{-1}{2} & \frac{1}{4} \end{bmatrix}$$

Solution : The eigenvalues are $\lambda = 5, 1, 1$.

The basis of the eigenspace corresponding $\lambda = 5$ is $\begin{bmatrix} 1 \\ 1 \\ 1 \end{bmatrix}$.

The basis of the eigenspace corresponding to $\lambda = 1$, is

$$\left\{ \begin{bmatrix} 1 \\ 0 \\ -1 \end{bmatrix}, \begin{bmatrix} 2 \\ -1 \\ 0 \end{bmatrix} \right\}.$$

2.5 Eigen Vectors and Linear Transformations

In this section, we will try to understand the matrix factorization $A = PDP^{-1}$ as a statement about linear transformations. If A is $m \times n$ matrix, then transformation T from $\mathbb{R}^n \to \mathbb{R}^m$ given by $\vec{x} \to A\vec{x}$. In fact, every linear transformation from $\mathbb{R}^n$ to $\mathbb{R}^m$ is a left multiplication by a matrix A, called the **standard matrix of T.** We will see the same sort of representation of any linear transformation between two finite dimensional vector spaces.

The Matrix of a Linear Transformation :

Let dimension of a vector space V be n and that of U be m, and let T be any linear transformation from V to U. Choose the ordered bases B and B' for V and U, respectively.

Given any $\vec{x} \in V$, the co-ordinate vector $[\vec{x}]_\beta$ is in $\mathbb{R}^n$ and the co-ordinate vector of its image $T(\vec{x})$, $[T(\vec{x})]_{\beta'}$, is in $\mathbb{R}^m$, see the Fig. 2.3.

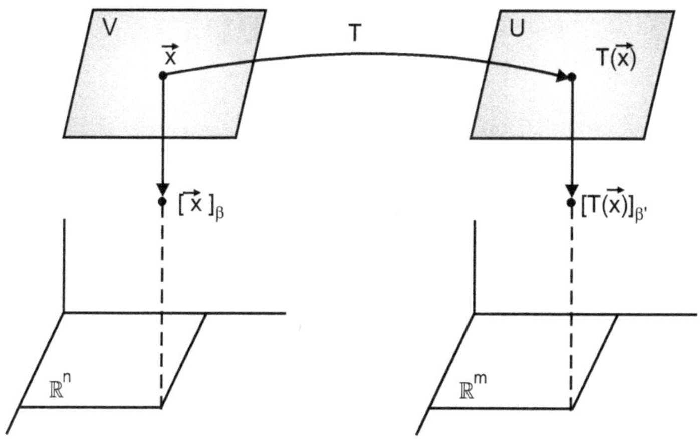

Fig. 2.3

A linear transformation from V to U.

Coordinate map from V to $\mathbb{R}^n$ and Co-ordinate map from U to $\mathbb{R}^m$.

We find the connection between $[\vec{x}]_\beta$ and $[T(\vec{x})]_{\beta'}$.

Let $\beta = \{\vec{v}_1, \vec{v}_2, ..., \vec{v}_n\}$ be a basis B for V. Then for any $\vec{x}$ is V, we have $\vec{x} = \alpha_1 \vec{v}_1 + \alpha_2 \vec{v}_2 + ... + \alpha_n \vec{v}_n$, then

$$[\vec{x}]_\beta = \begin{bmatrix} \alpha_1 \\ \alpha_2 \\ \vdots \\ \alpha_n \end{bmatrix}$$

and
$$T(\vec{x}) = T(\alpha_1 \vec{v}_1 + \alpha_2 \vec{v}_2 + ... + \alpha_n \vec{v}_n)$$
$$= \alpha_1 T(\vec{v}_1) + \alpha_2 T(\vec{v}_2) + ... + \alpha_n T(\vec{v}_n) \quad ... (1)$$

because T is a linear.

Now since β' is basis for U, we can write :

(1) In terms β' co-ordinate vectors :

$$[T(\vec{x})]_{\beta'} = \alpha_1 [T(\vec{v}_1)]_{\beta'} + ... + \alpha_n [T(\vec{v}_n)]_{\beta'} \quad ... (2)$$

Since, the β' coordinate vectors are in $\mathbb{R}^m$, the vector equation (2) can be written as a matrix equation, namely,

$$[T(\vec{x})]_{\beta'} = M[\vec{x}]_\beta \quad ... (3)$$

where, $$M = [T(\vec{v}_1)]_{\beta'} \; [T(\vec{v}_2)]_{\beta'} \; ... \; [T(\vec{v}_n)]_{\beta'}] \quad ... (4)$$

The matrix M is a matrix representation of T, called the **matrix for T relative** to the bases B and B'. See the Fig. 2.4.

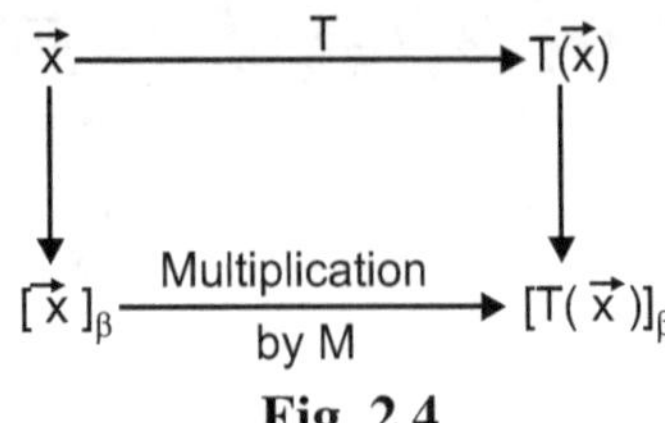

Fig. 2.4

Note : Equation (3) says that, so far as co-ordinate vectors are concerned, the action of T on $\vec{x}$ may be viewed as left multiplication by M.

Linear Transformation from V into V :

In general, we come across with the linear transformation from a vector space V into V itself, so that U = V and B' = B. In this case, the matrix M in (4) above is called the **matrix for T relative to the basis B,** or simply the **B-matrix for T** and is denoted by $[T]_B$. The B-matrix for $T : V \to V$ satisfies

$$[T(\vec{x})]_B = [T]_B \, [\vec{x}]_B, \quad \text{for all } \vec{x} \text{ in } V \qquad \ldots (5)$$

See the Fig. 2.5.

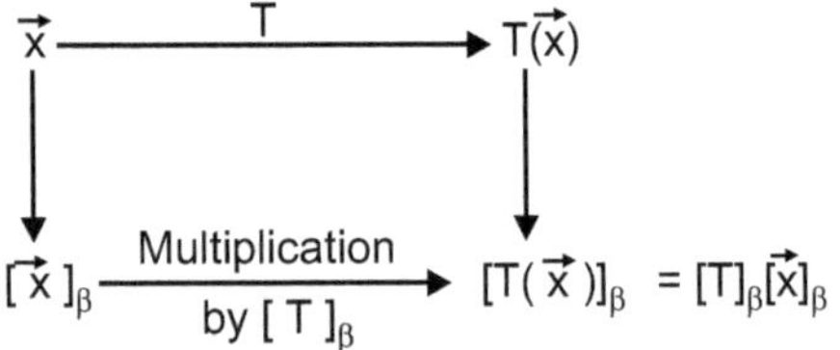

Fig. 2.5

Example 2.21 : *Let $B = \{\vec{u}_1, \vec{u}_2, \vec{u}_3\}$ and $B' = \{\vec{w}_1, \vec{w}_2\}$ be bases for vector spaces V and W, respectively. Let $T : V \to W$ be a linear transformation with the property that*

$$T(\vec{u}_1) = 3\vec{w}_1 - 5\vec{w}_2, \quad T(\vec{u}_2) = -\vec{w}_1 + 6\vec{w}_2, \quad T(\vec{u}_3) = 4\vec{w}_2.$$

Find the matrix for T relative to B and B'.

Solution : The B' co-ordinate vectors of the images of $\vec{u}_1$, $\vec{u}_2$ and $\vec{u}_3$ are given

$$[T(\vec{u}_1)]_{B'} = \begin{bmatrix} 3 \\ -5 \end{bmatrix}, \quad [T(\vec{u}_2)]_{B'} = \begin{bmatrix} -1 \\ 6 \end{bmatrix}, \quad [T(\vec{u}_3)]_{B'} = \begin{bmatrix} 0 \\ 4 \end{bmatrix}. \text{ Hence,}$$

the matrix M for T relative to B and B' is given by

$$M = [T(\vec{u}_1)]_{B'} \; [T(\vec{u}_2)]_{B'} \; [T(\vec{u}_3)]$$

$$\text{or} \quad M = \begin{bmatrix} 3 & -1 & 0 \\ -5 & 6 & 4 \end{bmatrix}.$$

Example 2.22 : *Let $S = \{\vec{e}_1,\ \vec{e}_2,\ \vec{e}_3\}$ be the standard basis for $\mathbb{R}^3$ and $\ B = \{\vec{u}_1,\ \vec{u}_2,\ \vec{u}_3\}$ be a basis for a vector space V, and $T : \mathbb{R}^3 \to V$ be a linear transformation with the property that*

$$T(x_1, x_2, x_3) = (x_3 - x_2)\,\vec{u}_1 + (-x_1 - x_3)\,\vec{u}_2 + (x_1 - x_2)\,\vec{u}_3$$

(a) Compute $T(\vec{e}_1), T(\vec{e}_2), T(\vec{e}_3)$

(b) Compute $[T(\vec{e}_1)]_B,\ [T(\vec{e}_2)]_B,\ T[(\vec{e}_3)]_B$

(c) Find the matrix for T relative to S and B.

Solution : (a) We know that $\vec{e}_1 = \begin{bmatrix} 1 \\ 0 \\ 0 \end{bmatrix}$, $\vec{e}_2 = \begin{bmatrix} 0 \\ 1 \\ 0 \end{bmatrix}$ and $\vec{e}_3 = \begin{bmatrix} 0 \\ 0 \\ 1 \end{bmatrix}$ are

standard basis vectors for $\mathbb{R}^3$.

Now, we compute images of the basis vectors.

$$T(\vec{e}_1) = 0\vec{u}_1 + (-1)\,\vec{u}_2 + (1-0)\,\vec{u}_3 = 0\vec{u}_1 - 1\vec{u}_2 + \vec{u}_3$$

$$T(\vec{e}_2) = (0-1)\,\vec{u}_1 + (0-0)\,\vec{u}_2 + (0-1)\,\vec{u}_3$$

$$= (-1)\,\vec{u}_1 + 0\vec{u}_2 + (-1)\,\vec{u}_2$$

$$T(\vec{e}_3) = 1\vec{u}_1 + (-1)\,\vec{u}_2 + 0\vec{u}_3$$

(b) Thus, the B co-ordinate vectors of the images are

$$[T(\vec{e}_1)]_B = \begin{bmatrix} 0 \\ -1 \\ 1 \end{bmatrix},\ [T(\vec{e}_2)]_B = \begin{bmatrix} -1 \\ 0 \\ -1 \end{bmatrix},\ [T(\vec{e}_3)]_B = \begin{bmatrix} 1 \\ -1 \\ 0 \end{bmatrix}$$

(c) The matrix T relative to S to B is

$$M = \begin{bmatrix} 0 & -1 & 1 \\ -1 & 0 & -1 \\ 1 & -1 & 0 \end{bmatrix}.$$

Example 2.23 : *Assume the mapping $T : P_2 \to P_2$ defined by*
$$T(a_0 + a_1 t + a_2 t^2) = 3a_0 + (5a_0 - a_1)\,t + (4a_1 + a_2)\,t^2 \text{ is linear.}$$

(a) Find the matrix representation of T relative to the basis $B = \{1, t, t^2\}$.

(b) Verify that $[T(P)]_\beta = [T]_\beta\,[P]_\beta$ for each P in P_2.

Solution : Let us compute the images of the basis vectors : $1, t, t^2$

$$T(1) = 3 + 5t + 0t^2$$

$$T(t) = 0 + (-1)t + 4t^2$$

$$T(t^2) = 0 + 0t + t^2$$

So we have β-co-ordinate vectors of $T(1)$, $T(t)$, $T(t^2)$ are given by

$$[T(1)]_\beta = \begin{bmatrix} 3 \\ 5 \\ 0 \end{bmatrix}, \ [T(t)]_\beta = \begin{bmatrix} 0 \\ -1 \\ 4 \end{bmatrix} \text{ and } [T(t^2)]_\beta = \begin{bmatrix} 0 \\ 0 \\ 1 \end{bmatrix}$$

Thus, $$[T]_\beta = \begin{bmatrix} 3 & 0 & 0 \\ 5 & -1 & 0 \\ 0 & 4 & 1 \end{bmatrix}$$

(b) For any $P(t)$ in P_2, we have $P(t) = a_0 + a_1 t + a_2 t^2$, a_0, a_1, a_2 scalars.

Then $T(P(t)) = 3a_0 + (5a_0 - a_1)t + (4a_1 + a_2) t^2$.

Therefore, β co-ordinate of $T(P(t))$ is given by,

$$[T(P(t))]_\beta = \begin{bmatrix} 3a_0 \\ 5a_0 - a_1 \\ 4a_1 + a_2 \end{bmatrix} \qquad \ldots \text{(i)}$$

On the other hands,

$$[T]_\beta \, [P(t)]_\beta = \begin{bmatrix} 3 & 0 & 0 \\ 5 & -1 & 0 \\ 0 & 4 & 1 \end{bmatrix} \begin{bmatrix} a_0 \\ a_1 \\ a_2 \end{bmatrix}$$

$$= \begin{bmatrix} 3a_0 \\ 5a_0 - a_1 \\ 4a_1 + a_2 \end{bmatrix} \qquad \ldots \text{(ii)}$$

From (i) and (ii), we get

$$[T(P(t))]_\beta = [T]_\beta \, [P(t)]_\beta$$

See the following figure to visualize pictorially above verification (b).

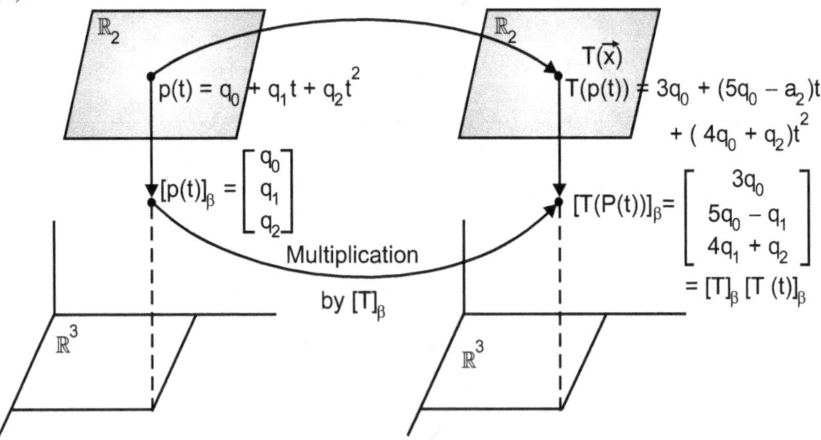

Fig. 2.6

Linear Transformation on $\mathbb{R}^n$

A linear transformation T from $\mathbb{R}^n$ into $\mathbb{R}^n$ appears as a matrix transformation, $\vec{x} \to A\vec{x}$. If A is diagonalizable, there is a basis β of $\mathbb{R}^n$ consisting of eigenvectors of A. The following theorem shows that, in this case, the β-matrix for T is diagonal. Diagonalising A amounts to finding a diagonal matrix representation of $\vec{x} \to A\vec{x}$.

| Theorem 9 | Diagonal Matrix Theorem

Suppose $A = PDP^{-1}$, where D is a diagonal $n \times n$ matrix. If β is the basis for $\mathbb{R}^n$ formed from columns of P, then D is the β-matrix for the transformation. $\vec{x} \to A\vec{x}$.

Proof : Let us denote the columns of P by $\vec{u}_1, \ldots, \vec{u}_n$, so that $\beta = \{\vec{u}_1, \ldots, \vec{u}_n\}$ and $P = [\vec{u}_1 \; \vec{u}_2 \ldots \vec{u}_n]$. In this case, P is the change of co-ordinates matrix P_β; where,

$$P[\vec{x}]_\beta = \vec{x} \text{ and } [\vec{x}]_\beta = P^{-1}\vec{x}$$

If $\quad T(\vec{x}) = A\vec{x} \quad$ for $\vec{x}$ in $\mathbb{R}^n$, then

$$
\begin{aligned}
[T]_B &= [[T(\vec{u}_1)]_B \ldots [T(\vec{u}_n)_\beta] && \text{By definition of } [T]_\beta \\
&= [[A\vec{u}_1]_\beta \ldots [A\vec{u}_n]_\beta] && \text{Since, } T(\vec{x}) = A\vec{x} \\
&= [P^{-1}A\vec{u}_1 \ldots P^{-1}A\vec{u}_n] && \text{By change of co-ordinate} \\
&= P^{-1}A[\vec{u}_1 \ldots \vec{u}_n] && \text{Matrix multiplication} \\
&= P^{-1}AP
\end{aligned}
$$

Since, $\quad\quad A = PDP^{-1}$, we have

$$[T]_\beta = P^{-1}AP = D$$

Example 2.24 : *Let $T : \mathbb{R}^2 \to \mathbb{R}^2$ be defined by $T(\vec{x}) = A\vec{x}$. Find a basis B for $\mathbb{R}^2$ with property that $[T]_B$ is diagonal, where $A = \begin{bmatrix} 5 & -3 \\ -7 & 1 \end{bmatrix}$.*

Solution : First we have to show that A is diagonalizable and find matrix P and a diagonal matrix D such that

$$A = PDP^{-1}.$$

Step 1 : To find eigenvalues of A. We have characteristic polynomial of A is

$$\det (A - \lambda I) = \det \begin{bmatrix} 5 - \lambda & -3 \\ -7 & 1 - \lambda \end{bmatrix}$$

$$= (5 - \lambda)(1 - \lambda) - 21$$

$$= \lambda^2 - 6\lambda + 5 - 21$$

$$= \lambda^2 - 6\lambda - 16$$

$$= (\lambda - 8)(\lambda + 2)$$

Thus, the eigenvalues of A are $\lambda = 8, -2$.

Step 2 : To find basis for eigenspaces for $\lambda = 8, -2$.

For $\lambda = 8$, we have to solve

$$(A - \lambda I)\,\vec{x} = \vec{0}$$

$$\begin{bmatrix} 5 - \lambda & -3 \\ -7 & 1 - \lambda \end{bmatrix} \begin{bmatrix} x_1 \\ x_2 \end{bmatrix} = \begin{bmatrix} 0 \\ 0 \end{bmatrix}$$

or
$$\begin{bmatrix} -3 & -3 \\ -7 & -7 \end{bmatrix} \begin{bmatrix} x_1 \\ x_2 \end{bmatrix} = \begin{bmatrix} 0 \\ 0 \end{bmatrix}, \quad \lambda = 8.$$

Clearly this is equivalent to $x_1 + x_2 = 0$ so $x_1 = -x_2$.

$$\therefore \qquad \begin{bmatrix} x_1 \\ x_2 \end{bmatrix} = \begin{bmatrix} -x_2 \\ x_2 \end{bmatrix} = x_2 \begin{bmatrix} -1 \\ 1 \end{bmatrix}$$

This shows that $\vec{u}_1 = \begin{bmatrix} -1 \\ 1 \end{bmatrix}$ is basis for the eigenspace corresponding to $\lambda = 8$.

Similarly, for $\lambda = -2$, we have

$$\begin{bmatrix} 7 & -3 \\ -7 & 3 \end{bmatrix} \begin{bmatrix} x_1 \\ x_2 \end{bmatrix} = \begin{bmatrix} 0 \\ 0 \end{bmatrix}$$

$$\begin{bmatrix} 7 & -3 \\ 0 & 0 \end{bmatrix} \begin{bmatrix} x_1 \\ x_2 \end{bmatrix} = \begin{bmatrix} 0 \\ 0 \end{bmatrix}$$

That is,
$$7x_1 - 3x_2 = 0$$

$$\therefore \qquad x_1 = \frac{3}{7} x_2$$

Thus, the general solution is

$$\begin{bmatrix} x_1 \\ x_2 \end{bmatrix} = \begin{bmatrix} \frac{3}{7} x_2 \\ x_2 \end{bmatrix} = x_2 \begin{bmatrix} \frac{3}{7} \\ 1 \end{bmatrix}$$

So taking $x_2 = 7$, we get, $\vec{u}_2 = \begin{bmatrix} 3 \\ 7 \end{bmatrix}$ – a basis for the eigenspace for corresponding to $\lambda = -2$.

Thus,
$$P = [\vec{u}_1 \ \vec{u}_2]$$

$$P = \begin{bmatrix} -1 & 3 \\ 1 & 7 \end{bmatrix}$$

and
$$D = \begin{bmatrix} 8 & 0 \\ 0 & -2 \end{bmatrix}$$

$\vec{u}_1$ and $\vec{u}_2$ are eigenvectors of A.

By theorem (8) D is the β-matrix for T where $\beta = \{\vec{u}_1, \vec{u}_2\}$. The mapping $\vec{x} \to A\vec{x}$ and $\vec{u} \to D\vec{u}$ describe the same linear transformation relative to different bases.

Similarity of Matrix Representations :

In the theorem 8, above we did not use the information that D was diagonal. Hence, if A is similar to a matrix C, with $A = PCP^{-1}$, then C is the β-matrix for the transformation $\vec{x} \to A\vec{x}$ when the basis B is formed from the columns of P.

Conversely, if $T : \mathbb{R}^n \to \mathbb{R}^n$ is defined by $T(\vec{x}) = A\vec{x}$, and if β is any basis for $\mathbb{R}^n$, then the β-matrix for T is similar to A. In fact, the calculations in theorem (8) show that if P is the matrix, whose columns come from the vectors in β, then $[T]_\beta = P^{-1}AP$. Thus, the set of all matrices similar to a matrix A coincides with the set of all matrix representations of the transformation $\vec{x} \to A\vec{x}$.

Example 2.25 : *Let $B = \{\vec{u}_1, \vec{u}_2, \vec{u}_3\}$ be a basis for a vector space V. Find $T(3\vec{u}_1 - 4\vec{u}_2)$ when T is a linear transformation from V to V, whose matrix relative β is,*

$$[T]_\beta = \begin{bmatrix} 0 & -6 & 1 \\ 0 & 5 & -1 \\ 1 & -2 & 7 \end{bmatrix}$$

Solution : We have,

$$T(3\vec{u}_1 - 4\vec{u}_2) = [T]_\beta \ [3\vec{u}_1 - 4\vec{u}_2]_\beta$$

$$= [T]_\beta \begin{bmatrix} 3 \\ -4 \\ 0 \end{bmatrix}_B$$

$$= \begin{bmatrix} 0 & -6 & 1 \\ 0 & 5 & -1 \\ 1 & -2 & 7 \end{bmatrix} \begin{bmatrix} 3 \\ -4 \\ 0 \end{bmatrix} = \begin{bmatrix} 24 \\ -20 \\ 11 \end{bmatrix}$$

Example 2.26 : *If $B = P^{-1}AP$ and $\vec{x}$ is an eigenvector of A corresponding to an eigenvalue λ, then show that $P^{-1}\vec{x}$ is an eigenvector of B corresponding also to λ.*

Solution : We have,

$$B(P^{-1}\vec{x}) = (P^{-1}AP)(P^{-1}\vec{x}) \qquad \because B = P^{-1}AP$$

$$= (P^{-1}A)(PP^{-1})\vec{x}$$

$$= (P^{-1}A) I \vec{x}$$

$$= P^{-1}A \vec{x}$$

$$= P^{-1}(\lambda\vec{x}) \qquad \because A\vec{x} = \lambda\vec{x}$$

$$= \lambda(P^{-1}\vec{x})$$

$\therefore \quad P^{-1}\vec{x}$ is an eigenvector of B corresponding to λ.

Example 2.27 : *Let A and B be $n \times n$ matrices. Then we say "A is similar to B" if there exists invertible matrix P such that $B = P^{-1}AP$. Then prove the following : A, B, C are $n \times n$ matrices.*

(a) A is similar to A.

(b) If A is similar to B, then B is similar to A.

(c) If A is similar to B and B is similar to C, then A is similar to C.

Solution : (a) Since, $A = I^{-1}AI$, I-identity matrix, A is similar to A.

(b) A is similar to B, so there is invertible matrix P such that $B = P^{-1}AP$. This equality premultiply by P and post multiply by P^{-1} gives

$$PBP^{-1} = P(P^{-1}AP) P^{-1} = (PP^{-1}) A (PP^{-1}) = A$$

That is, $A = PBP^{-1} = Q^{-1}BQ$, where $Q = P^{-1}$, and P^{-1} is invertible. Hence, B is similar to A.

(c) If A is similar to B, then there is invertible matrix P such that

$$B = P^{-1}AP \qquad \qquad \text{... (i)}$$

Also B is similar to C, then there exists an invertible matrix Q, such that

$$C = Q^{-1}BQ$$
$$= Q^{-1}(P^{-1}AP)\,Q \qquad \text{from equation (i)}$$
$$= (Q^{-1}P^{-1})\,A\,(PQ)$$
$$= (PQ)^{-1}\,A\,(PQ)$$

Since, P and Q are invertible, PQ is invertible, and thus A is similar to C.

Think Over It

1. If A is similar to B, then A^2 is similar to B^2.

2. If B is similar to A and A is similar to C, then B is similar to C.

3. Construct a non-diagonal 2×2 matrix that is diagonalizable but not invertible.

4. If A is square matrix then A and A^t have the same characteristic polynomial.

5. If A^2 is zero matrix, then the only eigenvalue of A is zero.

Summary

1. Eigenvalues and eigenvectors of a square matrix, computations of the same. eigenspace.

2. Eigenvectors corresponding to distinct eigenvalues.

3. **Characteristic equation :** The Invertible Matrix Theorem, similar matrices.

4. **Diagonalization of a matrix :** The Diagonalizable Theorem. Finding eigenvalues, eigenvectors of a matrix and checking the matrix is diagonalizable.

5. **Eigenvectors and linear transformation :** Matrix of a linear transformation. Linear transformations from V into V. Diagonal Matrix Representation.

Exercise

[A] Say True or False : Justify !

1. For n × n matrix A, any $\vec{x}$ in $\mathbb{R}^n$ is an eigenvector of A.

2. A scalar zero cannot be an eigenvalue of a matrix.

3. If $A = \begin{bmatrix} 1 & 6 \\ 5 & 2 \end{bmatrix}$ and $\vec{u} = \begin{bmatrix} 6 \\ -5 \end{bmatrix}$, then $\vec{u}$ is an eigenvector of A.

4. If $\vec{u}_1$ and $\vec{u}_2$ are linearly independent eigenvectors of a matrix, then they must correspond to two distinct eigenvalues.

5. An eigenspace of A is null space of A.

6. Similar matrices (n × n) have the same characteristic polynomials.

7. If A is n × matrix then the determinant of A is the product of the diagonal entries of A.

8. A is n × n matrix, then $\det A = (-1) \det A^t$.

9. In an n × n matrix row-replacement operation on A does not change the eigenvalues.

10. An n × n matrix A is diagonalizable, if $A = PDP^{-1}$ for some diagonal matrix D and some invertible matrix P.

11. An n × n matrix is diagonalizable if A has n-distinct eigenvalues.

[B] Multiple Choice Questions : Choose the Correct Alternative.

1. Define $T : P_2 \to \mathbb{R}^3$ by $T(P) = \begin{bmatrix} P(-1) \\ P(0) \\ P(1) \end{bmatrix}$. Then ……

   (a) $T(5 + 3t) = \begin{bmatrix} 2 \\ 5 \\ 8 \end{bmatrix}$

   (b) T is linear transformation, but $T(0) \neq 0$.

   (c) T is not linear transformation.

   (d) None of these

2. IF $B = P^{-1}AP$ and x is an eigenvector of A corresponding to an eigenvalue λ, then ……

   (a) $\vec{x}$ is also eigenvector of B corresponding to λ.

   (b) $P^{-1}\vec{x}$ is eigenvector of B corresponding to λ.

   (c) $P\vec{x}$ is eigenvector of B corresponding to λ.

   (d) None of these.

3. An $n \times n$ matrix A is diagonalizable, then ......
    (a) $\mathbb{R}^n$ has a basis of eigenvectors of A.
    (b) A is invertible.
    (c) A has n distinct eigenvalues.
    (d) None of the above.

4. Let A be 5×5 matrix, with two eigenvalues. A is diagonalizable, if ......
    (a) one eigenspace is of three dimensional and the other eigenspace is two dimensional.
    (b) one eigenspace of dimension 2 and the other of dimension one.
    (c) both eigenspace of dimension two each.
    (d) none of these.

5. If A is diagonalizable and invertible, then ......
    (a) A^{-1} is not diagonalizable.   (b) A^{-1} is also diagonalizable.
    (c) A^{-1} is diagonal.   (d) None of these

6. Let $A = \begin{bmatrix} 5 & -8 & 1 \\ 0 & 0 & 7 \\ 0 & 0 & -2 \end{bmatrix}$, then ......
    (a) A has three distinct eigenvalues.
    (b) A is not diagonalizable.
    (c) A is invertible.
    (d) None of these.

7. If $A = \begin{bmatrix} -1 & 0 \\ 0 & -1 \end{bmatrix}$, then A^5 is ......
    (a) $\begin{bmatrix} -1 & 0 \\ 0 & -1 \end{bmatrix}$ (b) $\begin{bmatrix} 1 & 0 \\ 0 & 1 \end{bmatrix}$
    (c) $\begin{bmatrix} -1 & -1 \\ 0 & 1 \end{bmatrix}$ (d) None of these

8. If $A = \begin{bmatrix} 5 & 3 \\ 3 & 5 \end{bmatrix}$, then the characteristic polynomial of A det $(A - \lambda I)$ is ......
    (a) $\lambda^2 - 10\lambda + 16$ (b) $\lambda^2 - 9\lambda + 9$
    (c) $\lambda^2 - 10\lambda + 25$ (d) None of these

9. If A and B are similar, then ......
    (a) det A = det B   (b) det A $\neq$ det B
    (c) det $(A - \lambda I) \neq$ det $(B - \lambda I)$ (d) None of these

10. $\vec{x}$ is eigenvector of A corresponding to λ, then $\vec{x}$ is eigenvector of A^3 corresponding to ......

    (a) λ (b) λ^2

    (c) λ^3 (d) None of these

[C] Theory Questions :

1. Prove that eigenvectors of a matrix A corresponding to distinct eigenvalues are linearly independent.

2. Let λ be an eigenvalue of an invertible matrix A, then show that λ^{-1} is an eigenvalue of A^{-1}.

3. If an n × n matrices A and B are similar, then prove that they have the same characteristic polynomial.

4. Prove that an n × n matrix A is diagonalizable if and only if A has n linearly independent eigenvectors.

5. Prove that an n × n matrix with n distinct eigenvalues is diagonalizable.

6. State and prove Diagonal Matrix Theorem.

[D] Numerical Problems :

1. Let $A = \begin{bmatrix} 1 & 6 \\ 5 & 2 \end{bmatrix}$, $\vec{u} = \begin{bmatrix} 6 \\ -5 \end{bmatrix}$ and $\vec{v} = \begin{bmatrix} 3 \\ -2 \end{bmatrix}$, are $\vec{u}$ and $\vec{v}$ are eigenvectors of A ?

2. Show that 7 is an eigenvalue of $\begin{bmatrix} 1 & 6 \\ 5 & 2 \end{bmatrix}$.

3. $A = \begin{bmatrix} 4 & -1 & 6 \\ 2 & 1 & 6 \\ 2 & -1 & 8 \end{bmatrix}$. Show that 2 is an eigenvalue of A and find the basis for the corresponding eigenspace.

4. Is $\begin{bmatrix} 4 \\ -3 \\ 1 \end{bmatrix}$ a vector of $\begin{bmatrix} 3 & 7 & 9 \\ -4 & -5 & 1 \\ 2 & 4 & 4 \end{bmatrix}$?

    If yes, find the corresponding eigenvalue.

5. Is $\lambda = 4$ an eigenvalue of $\begin{bmatrix} 3 & 0 & -1 \\ 2 & 3 & 1 \\ -3 & 4 & 5 \end{bmatrix}$?

    If so, find one corresponding eigenvector.

6. Find a basis for the eigenspace corresponding to the eigenvalue $\lambda = 3$, for $A = \begin{bmatrix} 4 & 2 & 3 \\ -1 & 1 & -3 \\ 2 & 4 & 9 \end{bmatrix}$.

7. Show that if A^2 is the zero matrix then the only eigenvalue of A is zero.

8. Find the characteristic equation of $A = \begin{bmatrix} 5 & -2 & 6 & -1 \\ 0 & 3 & -8 & 0 \\ 0 & 0 & 5 & 4 \\ 0 & 0 & 0 & 1 \end{bmatrix}$.

9. Find the characteristic polynomial and eigenvalues of $A = \begin{bmatrix} 2 & 1 \\ -1 & 4 \end{bmatrix}$.

10. Let $A = \begin{bmatrix} 7 & 2 \\ -4 & 1 \end{bmatrix}$, find A^k, $k \geq 1$ given that $A = PDP^{-1}$, where $P = \begin{bmatrix} 1 & 1 \\ -1 & -2 \end{bmatrix}$ and $D = \begin{bmatrix} 5 & 0 \\ 0 & 3 \end{bmatrix}$.

11. Diagonalize the matrix A, if possible, $A = \begin{bmatrix} 1 & 3 & 3 \\ -3 & -5 & -3 \\ 3 & 3 & 1 \end{bmatrix}$.

12. Determine whether A is diagonalizable $A = \begin{bmatrix} 2 & 4 & 3 \\ -4 & -6 & -3 \\ 3 & 3 & 1 \end{bmatrix}$.

13. Let A be a 4×4 matrix with eigenvalues 5, 3, and -2. Suppose that the eigenspace for $\lambda = 3$ is two dimensional. Is A diagonalizable ? Explain.

14. Let $A = \begin{bmatrix} 4 & 0 & -2 \\ 2 & 5 & 4 \\ 0 & 0 & 5 \end{bmatrix} = \begin{bmatrix} -2 & 0 & -1 \\ 0 & 1 & 2 \\ 1 & 0 & 0 \end{bmatrix} \begin{bmatrix} 5 & 0 & 0 \\ 0 & 5 & 0 \\ 0 & 0 & 4 \end{bmatrix} \begin{bmatrix} 0 & 0 & 1 \\ 2 & 1 & 4 \\ -1 & 0 & -2 \end{bmatrix}$ hold, then use the Diagonalization Theorem to find the eigenvalues of A and a basis for each eigenspace.

15. Determine whether the matrix, $A = \begin{bmatrix} 5 & -3 & 0 & 9 \\ 0 & 3 & 1 & -2 \\ 0 & 0 & 2 & 0 \\ 0 & 0 & 0 & 2 \end{bmatrix}$ is diagonalizable.

16. Let $B = \{\vec{u}_1, \vec{u}_2\}$ be basis for U and $\beta' = \{\vec{v}_1, \vec{v}_2, \vec{v}_3\}$ be a basis for V, let $T : U \to V$ be a linear transformation with the property that

$$T(\vec{u}_1) = 3\vec{v}_1 - 2\vec{v}_2 + 5\vec{v}_3 \text{ and } T(\vec{u}_2) = 4\vec{v}_1 + 7\vec{v}_2 - \vec{v}_3.$$

Find the matrix M for T relative to B and B'.

17. Let $T : P_2 \to P_2$ be defined by $T(a_0 + a_1t + a_2t^2) = a_1 + 2a_2t$ is a linear transformation.

    (a) Find the β-matrix for T, when B is the basis $\{1, t, t^2\}$.

    (b) Verify that $[T(P)]_B = [T]_B [P]_B$ for any $P \in P_2$.

18. Define $T : \mathbb{R}^2 \to \mathbb{R}^2$ by $T(\vec{x}) = A\vec{x}$, where $A = \begin{bmatrix} 7 & 2 \\ -4 & 1 \end{bmatrix}$. Find a basis β with the property that β matrix for T is a diagonal matrix.

19. Find $T(a_0 + a_1t + a_2t^2)$, if T is the linear transformation from P_2 to P_2 whose matrix relative $B = \{1, t, t^2\}$ is $[T]_\beta = \begin{bmatrix} 3 & 4 & 0 \\ 0 & 5 & -1 \\ 1 & -2 & 7 \end{bmatrix}$.

20. Let $T : P_2 \to P_3$ be the transformation that maps a polynomial $P(t) = a_0 + a_1t + a_2t^2$ into the polynomial $(t + 5) P(t)$. Then :

    (a) Find the image of $P(t) = 2 - t + t^2$.

    (b) Show that T is a linear transformation.

    (c) Find the matrix for T relative to the bases $\{1, t, t^2\}$ and $\{1, t, t^2, t^3\}$.

Answers

[A] (1) False   (2) False   (3) True   (4) False   (5) True   (6) True
    (7) False   (8) False   (9) True   (10) True   (11) True

[B] (1) - (a)   (2) - (b)   (3) - (a)   (4) - (a)   (5) - (b)   (6) - (a)   (7) - (a)
    (8) - (a)   (9) - (a)   (10) - (c)

[D] 1. $\vec{u}$ is an eigenvector but not $\vec{v}$.

    3. $\left\{ \begin{bmatrix} 1 \\ 2 \\ 0 \end{bmatrix}, \begin{bmatrix} -3 \\ 0 \\ 1 \end{bmatrix} \right\}$ is basis for eigenspace for $\lambda = 2$.

    4.   Yes, $\lambda = 0$

5.   Yes, $\begin{bmatrix} 1 \\ 1 \\ -1 \end{bmatrix}$

6. $\begin{bmatrix} -2 \\ 1 \\ 0 \end{bmatrix}, \begin{bmatrix} -3 \\ 0 \\ 1 \end{bmatrix}$

8. $(\lambda - 5)^2 (\lambda - 3)(\lambda - 1) = 0$ or $\lambda^4 - 14\lambda^3 + 68\lambda^2 - 130\lambda + 75 = 0$

9. $\lambda^2 - 6\lambda + 9, \ \lambda = 3, 3.$

10. $A^k = \begin{bmatrix} 2 \cdot 5^k - 3^k & 5^k - 3^k \\ 2 \cdot 3^k - 2 \cdot 5^k & 2 \cdot 3^k - 5^k \end{bmatrix}$

11.   A is diagonalizable with

$$P = \begin{bmatrix} 1 & -1 & -1 \\ -1 & 1 & 0 \\ 1 & 0 & 1 \end{bmatrix} \text{ and } D = \begin{bmatrix} 1 & 0 & 0 \\ 0 & -2 & 0 \\ 0 & 0 & -2 \end{bmatrix}.$$

12.   A is not diagonalizable as it has only two linearly independent eigenvectors.

13.   A is diagonalizable, because the eigenvectors corresponding to $\lambda = 5$ and $\lambda = -2$ along with eigenvectors for $\lambda = 3$ form a basis for $\mathbb{R}^4$.

14.   The eigenvalue of A are : $\lambda = 5, 5, 4$. The basis for the eigenspace corresponding to $\lambda = 5$ is $\left\{ \begin{bmatrix} -2 \\ 0 \\ 1 \end{bmatrix}, \begin{bmatrix} 0 \\ 1 \\ 0 \end{bmatrix} \right\}$ and basis for eigenspace corresponding to $\lambda = 4$ is $\begin{bmatrix} -1 \\ 2 \\ 0 \end{bmatrix}$.

15.   A is diagonalizable with

$$P = \begin{bmatrix} 1 & 3 & -1 & -1 \\ 0 & 2 & -1 & 2 \\ 0 & 0 & 1 & 0 \\ 0 & 0 & 0 & 1 \end{bmatrix} \text{ and } D = \begin{bmatrix} 5 & 0 & 0 & 0 \\ 0 & 3 & 0 & 0 \\ 0 & 0 & 2 & 0 \\ 0 & 0 & 0 & 2 \end{bmatrix}.$$

16. $M = \begin{bmatrix} 3 & 4 \\ -2 & 7 \\ 5 & -1 \end{bmatrix}$

17. $[T]_B = \begin{bmatrix} 0 & 1 & 0 \\ 0 & 0 & 2 \\ 0 & 0 & 0 \end{bmatrix}, \ [P]_B = \begin{bmatrix} a_0 \\ a_1 \\ a_2 \end{bmatrix} . \ [T(P)] = \begin{bmatrix} a_1 \\ 2a_2 \\ 0 \end{bmatrix}$

18. Eigenvalues of A are

$$\lambda = 5, 3 \begin{bmatrix} 1 \\ -1 \end{bmatrix} \text{ eigenvector corr} - \lambda = 5.$$

$$\begin{bmatrix} 1 \\ -2 \end{bmatrix} \text{ eigenvector corr} = 3.$$

$$\left\{ \begin{bmatrix} 1 \\ -1 \end{bmatrix}, \begin{bmatrix} 1 \\ -2 \end{bmatrix} \right\} \text{ is basis B for } \mathbb{R}^2.$$

19. $T(a_0 + a_1 t + a_2 t^2) = (3a_0 + 4a_1) + (5a_1 - a_2)\, t + (a_0 - 2a_1 + 7a_2)t^2$

20. (a) $T(P(t)) = 10 - 3t + 4t^2 + t^3.$

    (b) Verify for any two polynomials $p(t)$ and $q(t)$ and scalars, α, $T[p(t) + q(t)] = T(P(t)) + T(q(t))$ and

    $T[\alpha p(t)] = \alpha T(p(t))$

    (c) $\begin{bmatrix} 5 & 0 & 0 \\ 1 & 5 & 0 \\ 0 & 1 & 5 \\ 0 & 0 & 1 \end{bmatrix}.$

Chapter 3...

Inner Product on $\mathbb{R}^n$

Giuseppe Peano was an Italian mathematican, author of more than 200 books and papers. The first usage of the concept of a vector space with an inner product is due to him in 1898.

He was the cofounder of mathematical logic and set theory. The standard axiomatization of the natural numbers is named the "Peano Axioms" in his honor.

Giuseppe Peano

Introduction

In this chapter, we shall define general dot product alongwith its properties and applications to symmetric matrices. Although our geometric visualization does not extend beyond $\mathbb{R}^3$, interestingly it is possible to extend many familiar ideas beyond $\mathbb{R}^3$ for example, notion of distance between two vectors, length of a vector etc.

3.1 Inner Products

In the previous semester, we have dealt with the vectors and their properties. Now we turned towards the scalar quantity obtained from operations on vectors in $\mathbb{R}^n$.

Column Vector : A vector $\vec{u}$ in $\mathbb{R}^n$ which is of the form $n \times 1$ matrix is called column vector. i.e. column vector has n-rows and 1 - column.

For example, $\qquad \vec{u} = \begin{bmatrix} u_1 \\ u_2 \\ \vdots \\ u_n \end{bmatrix}_{n \times 1}$ is a column vector in $\mathbb{R}^n$.

Row Vector : A vector $\vec{v}$ in $\mathbb{R}^n$ which is of the form $1 \times n$ matrix is called row vector i.e. row vector in $\mathbb{R}^n$ has 1-row and n-column.

For example, $\qquad \vec{v} = [v_1 \ v_2 \ \dots \ v_n]_{1 \times n}$ is a row vector in $\mathbb{R}^n$.

Note that, the transpose of column vector is row vector while transpose of row vector is column vector.

The transpose of column vector $\vec{u}$ is denoted by $\vec{u}^t$ which is $1 \times n$ matrix. So that the matrix multiplication $\vec{u}^t \vec{v}$ is valid and it gives matrix of order 1×1, where $\vec{v}$ is $n \times 1$ column vector.

Inner Product : Given two vectors $\vec{u}, \vec{v}$ in $\mathbb{R}^n$ the number $\vec{u}^t \vec{v}$ is called as **inner product** of $\vec{u}$ and $\vec{v}$. It is also known as the dot product of two vectors $\vec{u}, \vec{v}$ and is denoted by $\vec{u} \cdot \vec{v}$.

Hence, if
$$\vec{u} = \begin{bmatrix} u_1 \\ u_2 \\ \vdots \\ u_n \end{bmatrix} \text{ and } \vec{v} = \begin{bmatrix} v_1 \\ v_2 \\ \vdots \\ v_n \end{bmatrix}$$

then the inner product of $\vec{u}$ and $\vec{v}$ is

$$\vec{u} \cdot \vec{v} = \vec{u}^t \vec{v} = [u_1 \ u_2 \ \ldots \ u_n]_{1 \times n} \begin{bmatrix} v_1 \\ v_2 \\ \vdots \\ v_n \end{bmatrix}_{n \times 1}$$

$$\Rightarrow \quad \vec{u}^t \vec{v} = [u_1 v_1 + u_2 v_2 + \ldots + u_n v_n]_{1 \times 1}$$

$$\vec{u}^t \vec{v} = \vec{u} \cdot \vec{v} = u_1 v_1 + u_2 v_2 + \ldots + u_n v_n$$

Example 3.1 : Let $\vec{u} = \begin{bmatrix} 1 \\ 2 \\ 3 \end{bmatrix}$ and $\vec{v} = \begin{bmatrix} -3 \\ -2 \\ -1 \end{bmatrix}$. Find $\vec{u} \cdot \vec{v}, \ \vec{v} \cdot \vec{u}$.

Solution : Given $\vec{u}^t = [1 \ 2 \ 3]_{1 \times 3}$ and $\vec{v} = \begin{bmatrix} -3 \\ -2 \\ -1 \end{bmatrix}_{3 \times 1}$

then
$$\vec{u} \cdot \vec{v} = \vec{u}^t \vec{v} = [1 \ 2 \ 3] \begin{bmatrix} -3 \\ -2 \\ -1 \end{bmatrix} = 1(-3) + 2(-2) + 3(-1)$$

$$= -10$$

$$\vec{v} \cdot \vec{u} = \vec{v}^t \vec{u} = [-3 \ -2 \ -1] \begin{bmatrix} 1 \\ 2 \\ 3 \end{bmatrix}$$

$$= (-3)(1) + (-2)(2) + (-1)(3) = -10$$

From the above example, it should be noted that the inner product or dot product of two vectors $\vec{u} \cdot \vec{v}$ as well as $\vec{v} \cdot \vec{u}$ are same i.e. in other words the dot product is commutative. Geometrically it gives relative orientation of two vectors.

| Theorem 1 | **Properties of dot product / inner product :**

If $\vec{u}, \vec{v}$ and $\vec{w}$ are any vectors in $\mathbb{R}^n$ and α be any scalar. Then :

(1) $\vec{u} \cdot \vec{v} = \vec{v} \cdot \vec{u}$ ... (Commutativity)

(2) $(\vec{u} + \vec{v}) \cdot \vec{w} = \vec{u} \cdot \vec{w} + \vec{v} \cdot \vec{w}$ ... (Distributivity)

(3) $(\alpha\vec{u}) \cdot \vec{v} = \alpha(\vec{u} \cdot \vec{v}) = \vec{u}(\alpha \cdot \vec{v})$ ... (Associtivity of scalar)

(4) $\vec{u} \cdot \vec{u} \geq 0$ ... (Non-negativity)

(5) $\vec{u} \cdot \vec{v} = 0 \Leftrightarrow \vec{u} = \vec{0}$

Proof : (1) Commutativity :

Let $\vec{u} = [u_1 \ u_2 \ ... \ u_n]$ and $\vec{v} = [v_1 \ v_2 \ ... \ v_n]$ be any vectors in $\mathbb{R}^n$.

Now, $\vec{u} \cdot \vec{v} = u_1v_1 + u_2v_2 + ... + u_nv_n$

$$\qquad\qquad ... \text{ By definition of dot product}$$

$$= v_1u_1 + v_2u_2 + ... + v_nu_n \qquad ... \text{ Property of scalars}$$

$$= \vec{v} \cdot \vec{u} \qquad ... \text{ By definition of dot product}$$

$$\therefore \quad \vec{u} \cdot \vec{v} = \vec{u} \cdot \vec{v}$$

Hence, dot product is commutative. ∎

(2) Distributive law :

Let $\vec{u} = [u_1 \ u_2 \ ... \ u_n]$, $\vec{v} = [v_1 \ v_2 \ ... \ v_n]$ and $\vec{w} = [w_1 \ w_2 \ ... \ w_n]$ be any vectors in $\mathbb{R}^n$. Now,

$$(\vec{u} + \vec{v}) \cdot \vec{w} = ([u_1 \ u_2 \ ... \ u_n] + [v_1 \ v_2 \ ... \ v_n] \cdot \vec{w}$$

$$= ([u_1 + v_1 \ u_2 + v_2 \ ... \ u_n + v_n]) \cdot [w_1 \ w_2 \ ... \ w_n]$$

$$\qquad\qquad ... \text{ Additive property of vectors}$$

$$= (v_1 + v_1) \, w_1 + (u_2 + v_2) \, w_2 + ... + (u_n + v_n) \, w_n$$

$$\qquad\qquad ... \text{ Definition of dot product}$$

$$= (u_1w_1 + v_1w_1) + (u_2w_2 + v_2w_2) + ... +$$

$$(u_nw_n + v_nw_n) \quad ... \text{ Distributive property of scalars}$$

$$= (u_1w_1 + u_2w_2 + \ldots + u_nw_n) +$$
$$(v_1w_1 + v_2w_2 + \ldots + v_nw_n) \quad \ldots \text{Property of scalars}$$
$$= \vec{u} \cdot \vec{w} + \vec{v} \cdot \vec{w}$$
$$\therefore \ (\vec{u} + \vec{v}) \cdot \vec{w} = \vec{u} \cdot \vec{w} + \vec{v} \cdot \vec{w}$$

Hence, dot product satisfies distributive property.

(3) Associativity of scalars :

Let α be any scalar, $\vec{u} = [u_1 \ u_2 \ \ldots \ u_n]$, $\vec{v} = [v_1 \ v_2 \ \ldots \ v_n]$ are any two vectors in $\mathbb{R}^n$.

$$\text{Now,} \ (\alpha\vec{u}) \cdot \vec{v} = (\alpha[u_1 \ u_2 \ \ldots \ u_n]) \cdot \vec{v}$$
$$= [\alpha u_1 \ \alpha u_2 \ \ldots \ \alpha u_n] \cdot \vec{v}$$
$$\ldots \text{Multiplication by scalar property}$$
$$= \alpha u_1 v_1 + \alpha u_2 v_2 + \ldots + \alpha u_n v_n$$
$$\ldots \text{Definition of dot product}$$
$$= \alpha(u_1 v_1 + u_2 v_2 + \ldots + u_n v_n)$$
$$= \alpha(\vec{u} \cdot \vec{v})$$
$$\therefore \quad (\alpha\vec{u}) \cdot \vec{v} = (\alpha\vec{u}) \cdot \vec{v}$$

Similarly, $\vec{u}(\alpha\vec{v}) = (\alpha\vec{u}) \ \vec{v}$

Hence, $(\alpha\vec{u}) \cdot \vec{v} = \alpha(\vec{u} \cdot \vec{v}) = \vec{u} \cdot (\alpha\vec{v})$

Hence, associativity holds.

(4) Non-negativity :

If $\vec{u} = [u_1 \ u_2 \ \ldots \ u_n]$ is any vector in $\mathbb{R}^n$. Then,

$$\vec{u} \cdot \vec{u} = u_1^2 + u_2^2 + \ldots + u_n^2$$

Now, the R.H.S. of the above equation is sum of squares, hence, it cannot be negative number

$$\therefore \qquad \vec{u} \cdot \vec{u} = u_1^2 + u_2^2 + \ldots + u_n^2 \geq 0$$

Hence, non-negativity property of dot product holds.

(5) If $\vec{u} = [u_1 \ u_2 \ \ldots \ u_n]$ be any vector in $\mathbb{R}^n$. Such that,

$$\vec{u} \cdot \vec{u} = 0.$$
$$\Leftrightarrow \ u_1^2 + u_2^2 + \ldots + u_n^2 = 0$$

$\Leftrightarrow \quad u_1 = 0, \ u_2 = 0, \ \ldots, \ u_n = 0$

$\ldots$ Sum of square is zero then each term is zero.

$\Leftrightarrow \quad \vec{u} = [0 \ \ 0 \ \ldots \ 0] = \vec{0}$

Hence, $\vec{u} \cdot \vec{u} = 0 \ \Leftrightarrow \ \vec{u} = \vec{0}$

Example 3.2 : If $\vec{u} = \begin{bmatrix} 1 \\ 2 \\ 3 \end{bmatrix}$, $\vec{v} = \begin{bmatrix} 2 \\ 3 \\ 4 \end{bmatrix}$ and $\vec{w} = \begin{bmatrix} 1 \\ 0 \\ 1 \end{bmatrix}$.

Find : $\dfrac{\vec{u} \cdot \vec{v}}{\vec{u} \cdot \vec{u}}, \ \dfrac{\vec{u} \cdot \vec{v}}{\vec{v} \cdot \vec{v}}, \ \left(\dfrac{\vec{u} \cdot \vec{v}}{\vec{u} \cdot \vec{u}}\right) \vec{w}, \ \dfrac{1}{\vec{u} \cdot \vec{u}} \, \vec{u}$

Solution : Given : $\vec{u} = \begin{bmatrix} 1 \\ 2 \\ 3 \end{bmatrix}$, $\vec{v} = \begin{bmatrix} 2 \\ 3 \\ 4 \end{bmatrix}$ and $\vec{w} = \begin{bmatrix} 1 \\ 0 \\ 1 \end{bmatrix}$

Step - I : $\vec{u} \cdot \vec{u} = [1 \ \ 2 \ \ 3] \begin{bmatrix} 1 \\ 2 \\ 3 \end{bmatrix} = 1 + 4 + 9 = 14$

$\qquad \vec{v} \cdot \vec{v} = [2 \ \ 3 \ \ 4] \begin{bmatrix} 2 \\ 3 \\ 4 \end{bmatrix} = 4 + 9 + 16 = 29$

$\qquad \vec{u} \cdot \vec{v} = [1 \ \ 2 \ \ 3] \begin{bmatrix} 2 \\ 3 \\ 4 \end{bmatrix} = 2 + 6 + 12 = 20$

Step - II : $\dfrac{\vec{u} \cdot \vec{v}}{\vec{u} \cdot \vec{u}} = \dfrac{20}{14} = \dfrac{10}{7} \Rightarrow \dfrac{\vec{u} \cdot \vec{v}}{\vec{u} \cdot \vec{u}} = \dfrac{10}{7}$

$\qquad \dfrac{\vec{u} \cdot \vec{v}}{\vec{v} \cdot \vec{v}} = \dfrac{20}{29}$

$\qquad \left(\dfrac{\vec{u} \cdot \vec{v}}{\vec{u} \cdot \vec{u}}\right) \vec{w} = \left(\dfrac{10}{7}\right) w = \dfrac{10}{7} \begin{bmatrix} 1 \\ 0 \\ 1 \end{bmatrix} = \begin{bmatrix} 10/7 \\ 0 \\ 10/7 \end{bmatrix}$

$\qquad \dfrac{1}{\vec{u} \cdot \vec{u}} \, \vec{u} = \dfrac{1}{14} \begin{bmatrix} 1 \\ 2 \\ 3 \end{bmatrix} = \begin{bmatrix} 1/14 \\ 2/14 \\ 3/14 \end{bmatrix} = \begin{bmatrix} 1/14 \\ 1/7 \\ 3/14 \end{bmatrix}$

3.1.1 Length of Vector

As we have seen the non-negative property of dot product i.e. for any vector $\vec{u}$ in $\mathbb{R}^n$ then $\vec{u} \cdot \vec{u} \geq 0$. So that one can thought of square root of it. We know that given any vector $\vec{u}$ with entries $u_1, u_2, \ldots, u_n$ it has magnitude as well as direction in $\mathbb{R}^n$.

The magnitude of vector $\vec{u} = [u_1\ u_2\ \ldots\ u_n]$ in $\mathbb{R}^n$ is nothing but length of vector $\vec{u}$, from origin $\vec{0}$ in $\mathbb{R}^n$.

Definition : The length (or norm) of $\vec{u}$ is the non-negative scalar quantity denoted by $\|\vec{u}\|$ and is defined as

$$\|\vec{u}\| = \sqrt{\vec{u} \cdot \vec{u}} = \sqrt{u_1^2 + u_2^2 + \ldots + u_n^2} \text{ and } \|\vec{u}\|^2 = \vec{u} \cdot \vec{u}$$

Geometrically, if $\vec{u} = (x, y)$ is any vector in $\mathbb{R}^2$ then $\|\vec{u}\| = \sqrt{x^2 + y^2}$ given by;

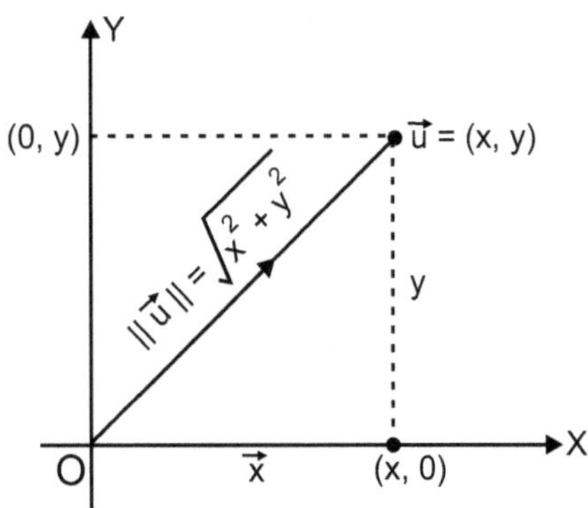

Fig. 3.1

The value $\|\vec{u}\| = \sqrt{x^2 + y^2}$ were calculated by using Pythagorean Theorem, applied to the diagonal of the rectangle. This definition can be extended to diagonal of the rectangular base in $\mathbb{R}^3$ and so on.

Properties of Length (or norm) :

$\boxed{\text{Theorem 2}}$ If α is any scalar and $\vec{u}$ is any vector in $\mathbb{R}^n$. Then :

(a) $\|\alpha\vec{u}\| = |\alpha|\ \|\vec{u}\|$

(b) If $\vec{u} \neq 0$ is any non-zero vector then $\dfrac{\vec{u}}{\|\vec{u}\|}$ has length 1 i.e.

$$\left\| \dfrac{\vec{u}}{\|\vec{u}\|} \right\| = 1 \text{ (called as unit vector)}$$

Proof : (a) If α is any scalar then by using the definition of norm,

$$\|\alpha\vec{u}\| = \sqrt{(\alpha\vec{u}) \cdot (\alpha\vec{u})}$$

Squaring both sides we get,

$$\|\alpha\vec{u}\|^2 = (\alpha\vec{u}) \cdot (\alpha\vec{u})$$

$$\Rightarrow \quad \|\alpha\vec{u}\|^2 = \alpha^2 \, (\vec{u} \cdot \vec{u}) \qquad \ldots \text{ Associativity of scalars}$$

$$\Rightarrow \quad \|\alpha\vec{u}\|^2 = \alpha^2 \, \|\vec{u}\|^2$$

Taking square root on both sides of the above equation we have,

$$\|\alpha\vec{u}\| = |\alpha| \, \|\vec{u}\|$$

(b) If $\vec{u} = [u_1 \; u_2 \; \ldots \; u_n]$ is any non-zero vector gives $\|\vec{u}\| > 0$.

Now, $\|\vec{u}\|$ is scalar, hence, $\left\| \dfrac{1}{\|\vec{u}\|} \vec{u} \right\| = \dfrac{1}{\|\vec{u}\|} \|\vec{u}\| \qquad \ldots \text{ From (a).}$

$$\Rightarrow \quad \left\| \dfrac{\vec{u}}{\|\vec{u}\|} \right\| = \dfrac{1}{\|\vec{u}\|} \|\vec{u}\| = 1 \qquad \ldots \; \| \, \|\vec{u}\| \, \| = \|\vec{u}\|$$

$$\therefore \quad \dfrac{\vec{u}}{\|\vec{u}\|} \text{ is non-zero vector of length 1 also called as unit vector. } \blacksquare$$

Note :

(1) A process of finding new vector $\vec{v} = \dfrac{1}{\|\vec{u}\|} \vec{u}$ which is of length 1

is also called normalizing $\vec{u}$, where $\vec{v}$ and $\vec{u}$ has same direction.

(2) A vector whose length is 1 is called unit vector i.e. for any non-zero vector $\vec{u}$ in $\mathbb{R}^n$, $\dfrac{1}{\|\vec{u}\|} \vec{u}$ is also non-zero vector whose length

(or norm) is 1.

(3) $\dfrac{-1}{\|\vec{u}\|}\,\vec{u}$ is unit vector whose direction is exactly opposite to $\vec{u}$ in $\mathbb{R}^n$, but with same magnitude.

Example 3.3 : *If $\vec{u} = (-1, 2, -1, 0)$. Find unit vector $\vec{v}_1$ in the same direction of u and unit vector $\vec{v}_2$ in the direction of $-\vec{u}$.*

Solution : Given : $\vec{u} = (-1, 2, -1, 0)$

$$\Rightarrow \quad \|\vec{u}\| = \sqrt{(-1)^2 + (2)^2 + (-1)^2 + 0^2} = \sqrt{6} \qquad \therefore \ \|\vec{u}\| = \sqrt{6}$$

Now, $\vec{v}_1 = \dfrac{1}{\|\vec{u}\|}\,\vec{u}$ while $\vec{v}_2 = \dfrac{-1}{\|\vec{u}\|}\,\vec{u}$

$$\therefore \quad \vec{v}_1 = \frac{1}{\sqrt{6}}(-1, 2, -1, 0) \quad \text{and} \quad \vec{v}_2 = \frac{-1}{\sqrt{6}}(-1, 2, -1, 0)$$

$$\therefore \quad \vec{v}_1 = \left(\frac{-1}{\sqrt{6}}, \frac{2}{\sqrt{6}}, \frac{-1}{\sqrt{6}}, 0\right) \quad \text{and} \quad \vec{v}_2 = \left(\frac{1}{\sqrt{6}}, \frac{-2}{\sqrt{6}}, \frac{1}{\sqrt{6}}, 0\right) \text{ with}$$

$$\|\vec{v}_1\|^2 = \left(\frac{-1}{\sqrt{6}}\right)^2 + \left(\frac{2}{\sqrt{6}}\right)^2 + \left(\frac{-1}{\sqrt{6}}\right)^2 + 0 = \frac{1}{6} + \frac{4}{6} + \frac{1}{6} = \frac{6}{6} = 1$$

and $\|\vec{v}_2\|^2 = \left(\frac{1}{\sqrt{6}}\right)^2 + \left(\frac{-2}{\sqrt{6}}\right)^2 + \left(\frac{1}{\sqrt{6}}\right)^2 + 0 = \frac{1}{6} + \frac{4}{6} + \frac{1}{6} = \frac{6}{6} = 1$

Hence, $\vec{v}_1$ and $\vec{v}_2$ both are unit vectors.

Geometrically, in $\mathbb{R}^2$, if $\vec{u} = (x, y)$ is vector in $\mathbb{R}^2$ then

$$\frac{\vec{u}}{\|\vec{u}\|} = \frac{(x, y)}{\sqrt{x^2 + y^2}} = \left(\frac{x}{\sqrt{x^2 + y^2}}, \frac{y}{\sqrt{x^2 + y^2}}\right) \text{ given by;}$$

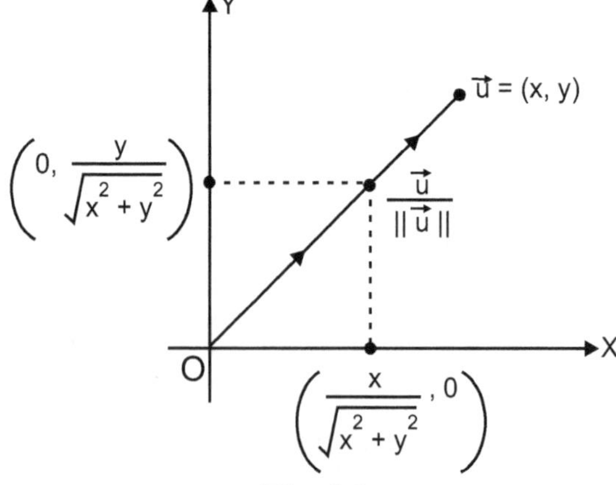

Fig. 3.2

3.1.2 Distance in $\mathbb{R}^n$

We know that on real line $\mathbb{R}$, for any two real numbers a and b the value $|a - b|$ or $|b - a|$ both are same and it is nothing but distance between a and b on $\mathbb{R}$.

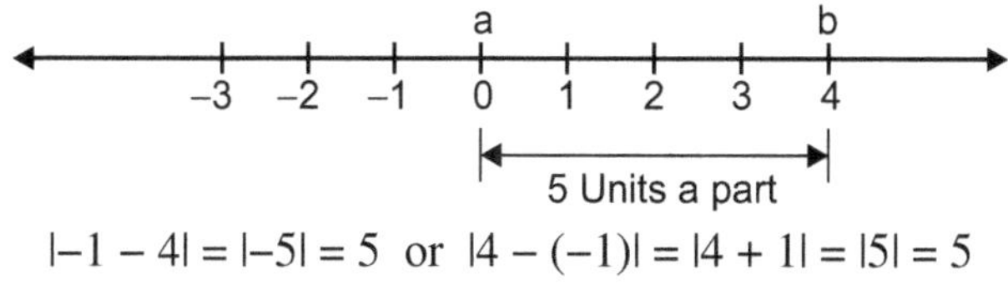

$$|-1 - 4| = |-5| = 5 \ \text{ or } \ |4 - (-1)| = |4 + 1| = |5| = 5$$

Fig. 3.3

The above definition can be extend to find distance between two vectors $\vec{u}$ and $\vec{v}$ in $\mathbb{R}^n$.

Distance between Two Vectors :

If $\vec{u}$ and $\vec{v}$ are vectors in $\mathbb{R}^n$ then the distance between $\vec{u}$ and $\vec{v}$ denoted by dist $(\vec{u}, \vec{v})$ and is given by the length of the vector $\vec{u} - \vec{v}$ i.e.

$$\text{dist}(\vec{u}, \vec{v}) = \|\vec{u} - \vec{v}\|$$

Note : The above definition of distance coincides with usual formula for Euclidean distance between two points in $\mathbb{R}^2$ or $\mathbb{R}^3$.

Example 3.4 : *If $\vec{u} = (u_1, u_2, \ldots, u_n)$ and $\vec{v} = (v_1, v_2, \ldots, v_n)$ are vectors $\mathbb{R}^n$ then find dist $(\vec{u}, \vec{v})$.*

Solution :

$$\text{dist}(\vec{u}, \vec{v}) = \|\vec{u} - \vec{v}\|$$
$$= \sqrt{(\vec{u} - \vec{v}) \cdot (\vec{u} - \vec{v})}$$
$$= \sqrt{(u_1 - v_1)^2 + (u_2 - v_2)^2 + \ldots + (u_n - v_n)^2}$$

Example 3.5 : *Compute the distance between two vectors*

$$\vec{u} = (1, 2, 3) \ \text{ and } \ \vec{v} = (3, 2, 1)$$

Solution : Given $\vec{u} = (1, 2, 3)$ and $\vec{v} = (3, 2, 1)$

$\Rightarrow \qquad \vec{u} - \vec{v} = (1 - 3, 2 - 2, 3 - 1) = (-2, 0, 2)$

$\Rightarrow \qquad \vec{u} - \vec{v} = (-2, 0, 2)$

$\therefore \qquad \text{dist}(\vec{u}, \vec{v}) = \|\vec{u} - \vec{v}\| = \sqrt{(-2)^2 + 0 + (2)^2} = \sqrt{8}$

$\Rightarrow \qquad \text{dist}(\vec{u}, \vec{v}) = \sqrt{8}$

Example 3.6 : Let $\vec{u} = \begin{bmatrix} 1 \\ 2 \\ 3 \end{bmatrix}$, $\vec{v} = \begin{bmatrix} 4 \\ 5 \\ 6 \end{bmatrix}$ find $\vec{u} \cdot \vec{v}$, $\|\vec{u}\|^2$, $\|\vec{v}\|^2$ and $\|\vec{u} + \vec{v}\|^2$.

Solution : Given : $\vec{u} = \begin{bmatrix} 1 \\ 2 \\ 3 \end{bmatrix}$ and $\vec{v} = \begin{bmatrix} 4 \\ 5 \\ 6 \end{bmatrix}$ then

$$\vec{u} \cdot \vec{v} = 4 + 10 + 18 = 32$$

$$\|\vec{u}\|^2 = (1)^2 + (2)^2 + (3)^2 = 1 + 4 + 9 = 14$$

$$\|\vec{v}\|^2 = (4)^2 + (5)^2 + (6)^2 = 16 + 25 + 36 = 76$$

$$\|\vec{u} + \vec{v}\|^2 = \|(5, 7, 9)\|^2 = (5)^2 + (7)^2 + (9)^2$$

$$= 25 + 49 + 81 = 155$$

$\therefore$　　$\|\vec{u}\|^2 = 14$, $\|\vec{v}\|^2 = 75$ and $\|\vec{u} + \vec{v}\|^2 = 155$

Remark : As the distance between $\vec{u}$ and $\vec{v}$ is given by $\|\vec{u} - \vec{v}\|$ which can be thought of as length of the vector $\vec{u} - \vec{v}$ and geometrically represented as

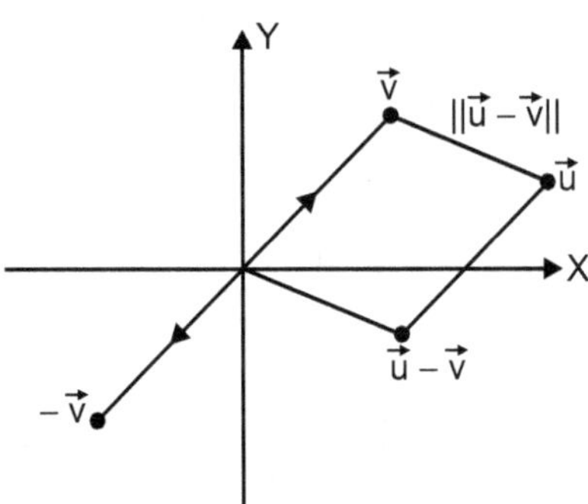

Fig. 3.4 : Distance between $\vec{u}$ and $\vec{v}$ is length of $\vec{u} - \vec{v}$

3.1.3 Orthogonal Vectors

We know that for two real numbers a and b whenever ab = 0, then either a = 0 or b = 0 or both. In the case of vectors whenever $\vec{u} \cdot \vec{v} = 0$ we cannot say it directly. For example, if $\vec{u} = (1, 0)$ and $\vec{v} = (0, 1)$ then $\vec{u} \cdot \vec{v} = 0 + 0 = 0$, where both $\vec{u}$ and $\vec{v}$ are non-zero vectors.

Definition : Orthogonal Vectors

The two vectors $\vec{u}$ and $\vec{v}$ in $\mathbb{R}^n$ are said to be orthogonal to each other if $\vec{u} \cdot \vec{v} = 0$.

Example 3.7 : Show that two vectors $\vec{u} = (1, 0, 1)$ and $\vec{v} = (-2, 0, 2)$ are orthogonal.

Solution : $\vec{u} \cdot \vec{v} = (1, 0, 1) \cdot (-2, 0, 2) = (-2) + 0 + (2) = 0$

$\Rightarrow \quad \vec{u} \cdot \vec{v} = 0$

Hence, $\vec{u}$ and $\vec{v}$ are orthogonal vectors.

Example 3.8 : *Show that two vectors $\vec{u}$ and $\vec{v}$ are orthogonal $\Leftrightarrow$ dist $(\vec{u}, \vec{v}) = $ dist $(\vec{u}, -\vec{v})$.*

Solution : Let $\vec{u}, \vec{v}$ are any two vectors in $\mathbb{R}^n$. Then :

$$[\text{dist} (\vec{u}, -\vec{v})]^2 = \|\vec{u} - (-\vec{v})\|^2$$

$$= \|\vec{u} + \vec{v}\|^2$$

$$= (\vec{u} + \vec{v}) \cdot (\vec{u} + \vec{v}) \ldots \text{ By definition of norm}$$

$$= \vec{u} \cdot \vec{u} + \vec{u} \cdot \vec{v} + \vec{v} \cdot \vec{u} + \vec{v} \cdot \vec{v}$$

$$\ldots \text{ Distributive law}$$

$$= \|\vec{u}\|^2 + \|\vec{v}\|^2 + 2\vec{u} \cdot \vec{v}$$

$$\ldots \text{ Commutativity of dot product}$$

$$\Rightarrow \quad [\text{dist} (\vec{u}, -\vec{v})]^2 = \|u\|^2 + \|v\|^2 + 2\vec{u} \cdot \vec{v} \qquad \ldots \text{(A)}$$

Now, $\quad [\text{dist} (\vec{u}, \vec{v})]^2 = \|\vec{u} - \vec{v}\|^2$

$$= (\vec{u} - \vec{v}) \cdot (\vec{u} - \vec{v}) \ldots \text{ By definition of norm}$$

$$= \vec{u} \cdot \vec{u} - \vec{u} \cdot \vec{v} - \vec{v} \cdot \vec{u} + \vec{v} \cdot \vec{v}$$

$$\ldots \text{ Distributive law}$$

$$= \|\vec{u}\|^2 + \|\vec{v}\|^2 - 2\vec{u} \cdot \vec{v}$$

$$\ldots \text{ Commutativity of dot product}$$

$$\Rightarrow \quad [\text{dist} (\vec{u}, \vec{v})]^2 = \|\vec{u}\|^2 + \|\vec{v}\|^2 - 2\vec{u} \cdot \vec{v} \qquad \ldots \text{(B)}$$

From equations (A) and (B),

$$[\text{dist}\,(\vec{u},\,\vec{v})]^2 \;=\; [\text{dist}\,(\vec{u}-\vec{v})]^2$$

$$\Leftrightarrow \quad \|\vec{u}\|^2 + \|\vec{v}\|^2 - 2\vec{u}\cdot\vec{v} \;=\; \|\vec{u}\|^2 + \|\vec{v}\|^2 + 2\vec{u}\cdot\vec{v}$$

$$\Leftrightarrow \qquad\qquad 4\vec{u}\cdot\vec{v} \;=\; 0$$

$$\Leftrightarrow \qquad\qquad \vec{u}\cdot\vec{v} \;=\; 0$$

$\Leftrightarrow \quad \vec{u}$ and $\vec{v}$ are perpendicular (orthogonal) to each other.

The above example can be stated as, $\|\vec{u}+\vec{v}\| = \|\vec{u}-\vec{v}\|$

$\Leftrightarrow \quad \vec{u}$ and $\vec{v}$ are orthogonal to each other, geometrically this is given in the Fig. 3.5.

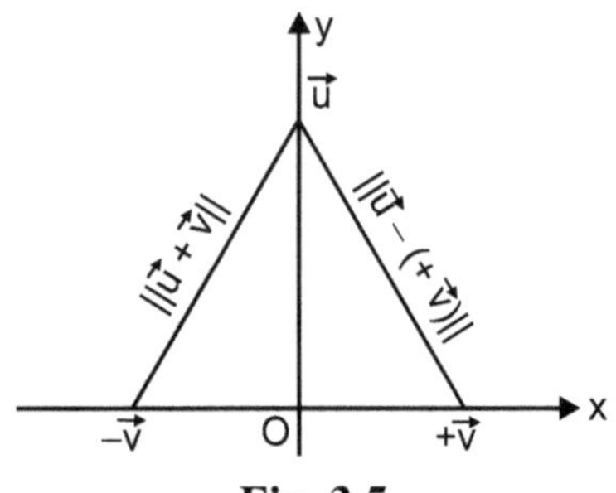

Fig. 3.5

Theorem 3 Two vectors $\vec{u}$ and $\vec{v}$ are orthogonal to each other if and only if $\|\vec{u}+\vec{v}\|^2 = \|\vec{u}\|^2 + \|\vec{v}\|^2$.

OR

Prove Pythogorean theorem for vectors in $\mathbb{R}^n$.

Proof : Let, $\vec{u}$ and $\vec{v}$ be any two vectors in $\mathbb{R}^n$. Now,

$$\|\vec{u}+\vec{v}\|^2 = \|\vec{u}\|^2 + \|\vec{v}\|^2$$

$\Leftrightarrow (\vec{u}+\vec{v})\cdot(\vec{u}+\vec{v}) = \vec{u}\cdot\vec{v} + \vec{v}\cdot\vec{v} \qquad\qquad \ldots$ Definition of norm

$\Leftrightarrow \vec{u}\cdot\vec{u} + \vec{u}\cdot\vec{v} + \vec{v}\cdot\vec{u} + \vec{v}\cdot\vec{v} = \vec{u}\cdot\vec{u} + \vec{v}\cdot\vec{v} \qquad \ldots$ Distributive law

$\Leftrightarrow 2\vec{u}\cdot\vec{v} = 0 \qquad\qquad\qquad\qquad \ldots \vec{u}\cdot\vec{v} = \vec{v}\cdot\vec{u}$

$\Leftrightarrow \vec{u}\cdot\vec{v} = 0$

$\Leftrightarrow \vec{u}$ and $\vec{v}$ are orthogonal to each other. ∎

Theorem 4 Verify the parallelogram law for vectors $\vec{u}, \vec{v}$ in $\mathbb{R}^n$ as

$$\|\vec{u} + \vec{v}\|^2 + \|\vec{u} - \vec{v}\|^2 = 2\|\vec{u}\|^2 + 2\|\vec{v}\|^2$$

Proof : Let $\vec{u}$ and $\vec{v}$ be any two vectors in $\mathbb{R}^n$. Now,

$$\|\vec{u} + \vec{v}\|^2 + \|\vec{u} - \vec{v}\|^2$$

$$= (\vec{u} + \vec{v}) \cdot (\vec{u} + \vec{v}) + (\vec{u} - \vec{v}) \cdot (\vec{u} - \vec{v}) \qquad \ldots \text{ Definition of norm}$$

$$= (\vec{u} \cdot \vec{u} + \vec{u} \cdot \vec{v} + \vec{v} \cdot \vec{u} + \vec{v} \cdot \vec{v}) + (\vec{u} \cdot \vec{u} - \vec{u} \cdot \vec{v} - \vec{v} \cdot \vec{u} + \vec{v} \cdot \vec{v})$$

$$\qquad\qquad \ldots \text{ Distributive law}$$

$$= 2\vec{u} \cdot \vec{u} + \vec{u} \cdot \vec{v} + \vec{u} \cdot \vec{v} + 2\vec{v} \cdot \vec{v} - \vec{u} \cdot \vec{v} - \vec{u} \cdot \vec{v}$$

$$\qquad\qquad \ldots \text{ Commutativity of dot product}$$

$$= 2\|\vec{u}\|^2 + 2\|\vec{v}\|^2 \qquad\qquad \ldots \text{ Definition of norm}$$

$$\therefore \quad \|\vec{u} + \vec{v}\|^2 + \|\vec{u} - \vec{v}\|^2 = 2\|\vec{u}\|^2 + 2\|\vec{v}\|^2 \qquad \blacksquare$$

Example 3.9 : *Let* $\vec{v} = \begin{bmatrix} a \\ b \end{bmatrix}$. *Describe the set H of vectors* $\begin{bmatrix} x \\ y \end{bmatrix}$ *that are orthogonal to* $\vec{v}$.

**Solution : Case - I : ** $H = \left\{ \begin{bmatrix} x \\ y \end{bmatrix} \middle| x, y \in \mathbb{R} \right\}$, if $\vec{v} = \begin{bmatrix} 0 \\ 0 \end{bmatrix}$ i.e. $a = 0$,

$b = 0$, then $\vec{v} \cdot \vec{u} = 0$ for all $\vec{u} \in H$.

$$\Rightarrow \quad [0 \ \ 0] \begin{bmatrix} x \\ y \end{bmatrix} = 0 \Leftrightarrow 0x + 0y = 0, \text{ for all } x, y \in \mathbb{R}$$

$$\Rightarrow \quad H = \mathbb{R}^2$$

Case - II : $\vec{v} = \begin{bmatrix} a \\ b \end{bmatrix} \neq 0$.

$$\therefore \quad \vec{v} \cdot \vec{u} = 0 \text{ for } \vec{u} \in H \Rightarrow [a \ \ b] \begin{bmatrix} x \\ y \end{bmatrix} = 0$$

$$\Rightarrow \quad ax + by = 0. \text{ Now if } a = 0 \text{ and } b \neq 0 \Rightarrow by = 0 \Rightarrow y = 0$$

$$\therefore \quad H = \left\{ \begin{bmatrix} x \\ y \end{bmatrix} \middle| y = 0 \right\} = \left\{ \begin{bmatrix} x \\ 0 \end{bmatrix} \middle| x \in \mathbb{R} \right\} = \text{X-axis in } \mathbb{R}^2$$

If $a \neq 0$ and $b = 0 \Rightarrow ax = 0 \Rightarrow x = 0$

$$\therefore \quad H = \left\{ \begin{bmatrix} x \\ y \end{bmatrix} \middle| x = 0 \right\} = \left\{ \begin{bmatrix} 0 \\ y \end{bmatrix} \middle| y \in \mathbb{R} \right\} = \text{Y-axis in } \mathbb{R}^2$$

If $a \neq 0,\ b \neq 0 \Rightarrow ax + by = 0$ gives $y = \left(\dfrac{-a}{b}\right) x$

$$\therefore \quad H = \left\{ \begin{bmatrix} x \\ y \end{bmatrix} \,\middle|\, x,\, y \in \mathbb{R} \right\} = \left\{ \begin{bmatrix} x \\ \left(\dfrac{-a}{b}\right) x \end{bmatrix} \,\middle|\, x \in \mathbb{R} \right\}$$

$$= \left\{ x \begin{bmatrix} 1 \\ \left(\dfrac{-a}{b}\right) \end{bmatrix} \,\middle|\, x \in \mathbb{R} \right\}$$

H is line passing through origin with slope $m = \left(\dfrac{-a}{b}\right)$.

Remark : In the above example, instead of finding a single vector one can find set of vectors which are orthogonal to a single vector. This process of finding set of orthogonal vectors to single vector is called **Orthogonal Complement.**

3.1.4 Orthogonal Complement

Suppose, we are standing straight on the floor and we ourselves can think of as single vector then all the vector "on the floor" are orthogonal to us. As we are perpendicular (orthogonal) to all the vectors on the floor.

The zero vector is orthogonal to every vector in $\mathbb{R}^n$ because $\vec{0} \cdot \vec{v} = 0,\ \forall\, \vec{v} \in \mathbb{R}^n$.

Definition : Orthogonal Complement

If a vector $\vec{u}$ is orthogonal to every vector in a subspace W of $\mathbb{R}^n$ then $\vec{u}$ is orthogonal to W.

The set of all vectors $\vec{u}$ that are orthogonal to W is called the orthogonal complement of W and is denoted by $W^{\perp}$. The symbol $W^{\perp}$ can be read as W per or W perpendicular.

Geometrically, this is shown in the following figure.

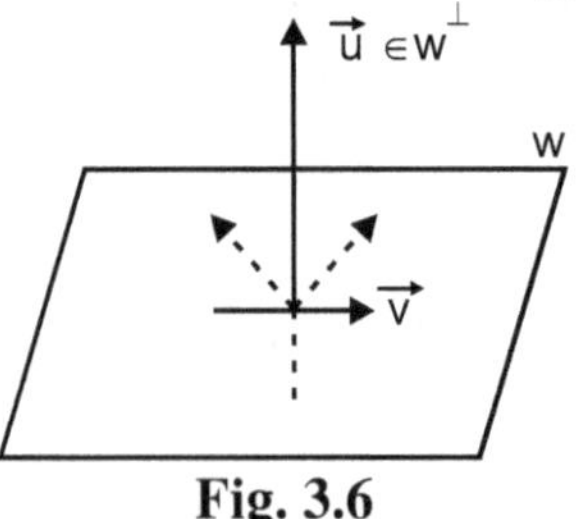

Fig. 3.6

Example 3.10 : If a vector $\vec{u}$ is orthogonal to $\vec{v}$ and $\vec{w}$ both then $\vec{u}$ is orthogonal to $\vec{v} + \vec{w}$.

Solution : If $\vec{u}$ is orthogonal to $\vec{v}$ and $\vec{w}$

$$\Rightarrow \quad \vec{u} \cdot \vec{v} = 0 \text{ and } \vec{u} \cdot \vec{w} = 0$$

$$\text{Now,} \quad \vec{u} \cdot (\vec{v} + \vec{w}) = \vec{u} \cdot \vec{v} + \vec{u} \cdot \vec{w} \qquad \text{... Distributive law}$$

$$= 0 + 0 \qquad \text{... } \vec{u} \cdot \vec{v} = 0 \text{ and } \vec{u} \cdot \vec{w} = 0$$

$$= 0 \qquad \text{... Property zero}$$

$$\Rightarrow \quad \vec{u} \cdot (\vec{v} + \vec{w}) = 0$$

Hence, $\vec{u}$ is orthogonal to $\vec{v} + \vec{w}$.

Remark : If $\vec{u}, \vec{v} \in W$ then we know that Span $\{\vec{u}, \vec{v}\}$ is nothing but set of all the vectors of the type $\alpha\vec{u} + \beta\vec{v}$ where α and β are scalars.

Example 3.11 : If $\vec{u}$ is orthogonal to $\vec{w}$ and $\vec{v}$ then show that $\vec{u}$ is orthogonal to every vector $\vec{z}$ in Span $\{\vec{v}, \vec{w}\}$.

Solution : Given that, $\vec{u}$ is orthogonal to $\vec{v}$ and $\vec{w}$.

$$\Rightarrow \quad \vec{u} \cdot \vec{v} = 0 \text{ and } \vec{u} \cdot \vec{w} = 0$$

Now, Span $\{\vec{v}, \vec{w}\} = \{\alpha\vec{u} + \beta\vec{w} \,/\, \alpha, \beta \text{ scalars}\}$

To show that, $\vec{u}$ is orthogonal to every vector in Span $\{\vec{v}, \vec{w}\}$.

Let $\vec{z} \in$ Span $\{\vec{v}, \vec{w}\} \Rightarrow \vec{z} = \alpha\vec{v} + \beta\vec{w}$ for some scalar α, β.

$$\text{Now,} \quad \vec{u} \cdot \vec{z} = \vec{u} \cdot (\alpha\vec{v} + \beta\vec{v})$$

$$= \vec{u} \cdot (\alpha\vec{v}) + \vec{u} \cdot (\beta\vec{v}) \qquad \text{... Distributive law}$$

$$= \alpha(\vec{u} \cdot \vec{v}) + \beta(\vec{u} \cdot \vec{v}) \qquad \text{... Associativity of scalar}$$

$$= \alpha 0 + \beta 0 \qquad \text{... } \vec{u} \cdot \vec{v} = 0 = \vec{u} \cdot \vec{w}$$

$$= 0$$

$$\Rightarrow \quad \vec{u} \cdot \vec{z} = 0 \Rightarrow \vec{u} \text{ is orthogonal to } \vec{z}.$$

$$\Rightarrow \quad \vec{u} \text{ is orthogonal to every vector in Span } \{\vec{v}, \vec{w}\}.$$

Example 3.12 : Let $W = \text{Span}\ \{\vec{v}_1, \vec{v}_2, ..., \vec{v}_p\}$. Show that if $\vec{u}$ is orthogonal to each vector $\vec{v}_j$, $1 \le j \le p$ then $\vec{u}$ is orthogonal to every vector in W.

Solution: Given, $\vec{u}$ is orthogonal to every vector $\vec{v}_j$, $1 \le j \le p$, where,

$$W = \text{Span}\ \{\vec{v}_1, \vec{v}_2, ..., \vec{v}_p\} \Rightarrow \vec{u} \cdot \vec{v}_j = 0,\ 1 \le j \le p$$

Let $\vec{w} \in \text{Span}\ \{\vec{v}_1, \vec{v}_2, ..., \vec{v}_p\} \Rightarrow \vec{w} = \alpha_1 \vec{v}_1 + \alpha_2 \vec{v}_2 + ... + \alpha_p \vec{v}_p$

$$
\begin{aligned}
\text{Now,} \quad \vec{u} \cdot \vec{w} &= \vec{u} \cdot (\alpha_1 \vec{v}_1 + \alpha_2 \vec{v}_2 + ... + \alpha_p \vec{v}_p\} \\
&= \vec{u} \cdot (\alpha_1 \vec{v}_1) + \vec{u} \cdot (\alpha_2 \vec{v}_2) + ... + \vec{u} \cdot (\alpha_p \vec{v}_p) \\
&\qquad\qquad\qquad\qquad\qquad\qquad\qquad\text{... Distributive law} \\
&= \alpha_1(\vec{u} \cdot \vec{v}_1) + \alpha_2(\vec{u} \cdot \vec{v}_2) + ... + \alpha_p(\vec{u} \cdot \vec{v}_p) \\
&\qquad\qquad\qquad\qquad\qquad\qquad\qquad\text{... Associativity law} \\
&= \alpha_1 0 + \alpha_2 0 + ... + \alpha_p 0 \qquad \because \vec{u} \cdot \vec{v}_j = 0\ 1 \le j \le p \\
&= 0
\end{aligned}
$$

$$\Rightarrow \quad \vec{u} \cdot \vec{w} = 0$$

$\Rightarrow \quad \vec{u}$ is orthogonal to every vector in W.

Theorem 4 (1) A vector u is in $W^\perp$ if and only if $\vec{u}$ is orthogonal to every vector in a set that spans W.

(2) $W^\perp$ is a subspace of $\mathbb{R}^n$.

Proof : If $\vec{u} \in W^\perp$, then $\vec{u}$ is orthogonal to every vector in W in particular $\vec{u}$ is orthogonal to every vector in the set S that spans W, as $S \subset W$. Conversely, if $\vec{u}$ is orthogonal to each vector in the set, that spans W then by above example 3.12, $\vec{u}$ is orthogonal to every vector in W, hence $\vec{u} \in W^\perp$.

Step I : If $\vec{u}, \vec{v} \in W^\perp$ and α, β are any scalars then for $\alpha\vec{u} + \beta\vec{v}$ is also orthogonal to every vector in W, because for any $\vec{w} \in W$.

$$
\begin{aligned}
\text{We have, } (\alpha\vec{u} + \beta\vec{v}) \cdot \vec{w} &= (\alpha\vec{u}) \cdot \vec{w} + (\beta\vec{v}) \cdot \vec{w} \qquad \text{... Distributive law} \\
&= \alpha(\vec{u} \cdot \vec{w}) + \beta(\vec{v} \cdot \vec{w}) \qquad \text{... Associativity} \\
&= 0 \qquad\qquad\qquad\qquad\qquad \text{... } \vec{u} \cdot \vec{w} = 0 = \vec{v} \cdot \vec{w}
\end{aligned}
$$

$$\therefore \qquad \alpha\vec{u} + \beta\vec{v} \in W^\perp$$

Hence, by definition of subspace, $W^\perp$ is subspace of $\mathbb{R}^n$.   ■

$\boxed{\textbf{Theorem 5}}$ Let A be an m × n matrix. The orthogonal complement of row space of A is the null space of A and the orthogonal complement of column space of A is the null space of A^t i.e.

$$(\text{Row } A)^\perp = \text{Nul } A \quad \text{and} \quad (\text{Col } A)^\perp = \text{Nul } A^t$$

Proof : If $x \in (\text{Row } A)^\perp \Rightarrow x$ is orthogonal to Row A i.e. x is orthogonal to each row of A.

$\Rightarrow \quad Ax = b$ has only solution $b = 0$

i.e. $Ax = 0$

$\Rightarrow \quad x \in \text{Nul } A$

$\therefore \quad (\text{Row } A)^\perp \subseteq \text{Nul } A \hspace{6cm} \ldots \text{(I)}$

Conversely, if $x \in \text{Nul } A \Rightarrow Ax = 0$

$\Rightarrow Ax = 0$ which is possible only if x is orthogonal to every row of A.

$\Rightarrow \quad x \in (\text{Row } A)^\perp$

$\therefore \quad \text{Nul } A \subseteq (\text{Row } A)^\perp \hspace{6cm} \ldots \text{(II)}$

$\therefore \quad$ From equations (I) and (II)

$\quad$ Nul $A = (\text{Row } A)^\perp$ Proved.

Now as A is any matrix the statement holds for A^t also.

$\therefore \quad (\text{Col } A)^\perp = (\text{Row } A^t)^\perp = \text{Nul } A^t.$ ■

3.2 Orthogonal Sets

In the previous section, we have seen the process of normalizing the non-zero vector, also under what conditions two vectors are orthogonal to each other. These two definitions when we combine; give rise to new kind of a vector called as orthonormal vector.

Definition : Orthogonal set of vectors

A set of vectors $S = \{\vec{u}_1, \vec{u}_2, \ldots, \vec{u}_p\}$ in $\mathbb{R}^n$ is said to be orthogonal if $\vec{u}_i \cdot \vec{u}_j = 0, \ i \neq j$

Definition : Orthonormal set of vectors

A set of vectors $S = \{\vec{v}_1, \vec{v}_2, \ldots, \vec{v}_k\}$ in $\mathbb{R}^n$ is said to be orthonormal if;

(a) $\|\vec{v}_j\| = 1, \ 1 \leq j \leq k$ (b) $\vec{v}_i \cdot \vec{v}_j = 0 \ \ i \neq j$

Thus, from the definition itself one can see easily that every orthonormal set is orthogonal set. Where the norm used is Euclidean norm.

Example 3.13 : Show that $S = \{\vec{u}_1, \vec{u}_2, \vec{u}_3\}$ is orthogonal but not orthonormal set, where $\vec{u}_1 = \begin{bmatrix} 1 \\ 0 \\ 0 \end{bmatrix}$, $\vec{u}_2 = \begin{bmatrix} 0 \\ 0 \\ 1 \end{bmatrix}$, $\vec{u}_3 = \begin{bmatrix} 0 \\ 0 \\ 2 \end{bmatrix}$.

Solution : $S = \left\{ \vec{u}_1 = \begin{bmatrix} 1 \\ 0 \\ 0 \end{bmatrix}, \vec{u}_2 = \begin{bmatrix} 0 \\ 1 \\ 0 \end{bmatrix}, \vec{u}_3 = \begin{bmatrix} 0 \\ 0 \\ 2 \end{bmatrix} \right\}$ then, we have

$$\vec{u}_1 \cdot \vec{u}_2 = 1(0) + 0(1) + 0(0) = 0, \quad \vec{u}_1 \cdot \vec{u}_3 = 1(0) + 0(0) + 0(2) = 0$$

$$\vec{u}_2 \cdot \vec{u}_3 = 0(0) + 1(0) + 0(2) = 0$$

Now, $\|\vec{u}_1\| = \sqrt{(1)^2 + (0)^2 + (0)^2} = 1$, $\|\vec{u}_2\| = \sqrt{(0)^2 + (1)^2 + (0)^2} = 1$

but $\quad \|\vec{u}_3\| = \sqrt{(0)^2 + (0)^2 + (2)^2} = 2 \quad \therefore \quad \|\vec{u}_3\| \neq 1$

hence, S is orthogonal but not orthonormal set.

Theorem 6 If $S = \{\vec{u}_1, \vec{u}_2, ..., \vec{u}_k\}$ is an orthogonal set of non-zero vectors in $\mathbb{R}^n$, then S is linearly independent and hence, S becomes basis for the subspace spanned by S.

Proof : Step I : Given that $S = \{\vec{u}_1, \vec{u}_2, ..., \vec{u}_k\}$ is an orthogonal set of non-zero vectors in $\mathbb{R}^n$, so $\vec{u}_i \cdot \vec{u}_j = 0$, for all $i \neq j$.

To show that, S is linearly independent it is equivalent to show that, whenever, $\alpha_1 \vec{u}_1 + \alpha_2 \vec{u}_2 + ... + \alpha_k \vec{u}_k = 0$ then $\alpha_1 = 0 = \alpha_2 = ... = \alpha_k$.

So consider, $\alpha_1 \vec{u}_1 + \alpha_2 \vec{u}_2 + ... + \alpha_k \vec{u}_k = \vec{0}$

taking dot product by $\vec{u}_1$ on both sides we get,

$$\vec{u}_1 \cdot [\alpha_1 \vec{u}_1 + \alpha_2 \vec{u}_2 + ... + \alpha_k \vec{u}_k] = \vec{u}_1 \cdot \vec{0}$$

$\Rightarrow \vec{u}_1 \cdot (\alpha_1 \vec{u}_1) + \vec{u}_1 \cdot (\alpha_2 \vec{u}_2) + ... + \vec{u}_1 \cdot (\alpha_k \vec{u}_k) = \vec{0}$ ... Distributivity

$\Rightarrow \quad \alpha_1(\vec{u}_1 \cdot \vec{u}_1) + \alpha_2(\vec{u}_1 \cdot \vec{u}_2) + ... + \alpha_k(\vec{u}_1 \cdot \vec{u}_k) = 0$

... Property of scalar

$\Rightarrow \quad \alpha_1 (\vec{u}_1 \cdot \vec{u}_1) = 0 \qquad\qquad \because \vec{u}_i \cdot \vec{u}_j = 0$ for all $i \neq j$

Now, as $\vec{u}_1$ is non-zero vector $\Rightarrow \vec{u}_1 \cdot \vec{u}_2 > 0 \quad \therefore \quad \alpha_1 = 0$

Similarly, α_2, α_3, ..., α_k all are zero. Hence, by definition S is linearly independent set.

Step II : Let W = Span S

$\Rightarrow$ S is linearly independent as well as S spans W.

$\therefore$ By the definition of basis; S becomes basis for W.

Hence, S becomes basis for the subspace spanned by S.   ∎

Remark : From the above theorem, one can add extra condition to basis to have new kind of basis.

Definition : Orthogonal basis

A basis S for the subspace W of $\mathbb{R}^n$ is said to be orthogonal basis if S is itself orthogonal.

As we have seen in the previous sections that, whenever

$S = \{\vec{u}_1, \vec{u}_2, ..., \vec{u}_k\}$ is basis for subspace W of $\mathbb{R}^n$. Then for any $\vec{w} \in S$ we have,

$$\vec{w} = \alpha_1 \vec{u}_1 + \alpha_2 \vec{u}_2 + ... + \alpha_k \vec{u}_k$$

where, $\alpha_1, \alpha_2, ..., \alpha_k$ are weights (scalars). If S is orthogonal basis then it is easy to calculate the weights by using formula which is obtained as follows :

Theorem 7 If $S = \{\vec{u}_1, \vec{u}_2, ..., \vec{u}_k\}$ is an orthogonal basis for a subspace W of $\mathbb{R}^n$. Then for $\vec{w} \in S$; $\vec{w} = \alpha_1 \vec{u}_1 + \alpha_2 \vec{u}_2 + ... + \alpha_k \vec{u}_k$

where, $\alpha_j = \dfrac{\vec{w} \cdot \vec{u}_j}{\vec{u}_j \cdot \vec{u}_j}$ $j = 1, 2, ..., k$

Proof : Since, $S = \{\vec{u}_1, \vec{u}_2, ..., \vec{u}_k\}$ is an orthogonal basis for a subspace W of $\mathbb{R}^n \Rightarrow W = \text{Span } S = \text{Span } \{\vec{u}_1, \vec{u}_2, ..., \vec{u}_k\}$

$\therefore$ For any $\vec{w} \in W \Rightarrow \vec{w} = \alpha_1 \vec{u}_1 + \alpha_2 \vec{u}_2 + ... + \alpha_k \vec{u}_k$

Now, $\vec{w} \cdot \vec{u}_j = (\alpha_1 \vec{u}_1 + ... + \alpha_{j-1} \vec{u}_{j-1} + \alpha_j \vec{u}_j + ... + \alpha_k \vec{u}_k) \cdot \vec{u}_j$

$\Rightarrow \vec{w} \cdot \vec{u}_j = (\alpha_1 \vec{u}_1) \cdot \vec{u}_j + ... + (\alpha_{j-1} \vec{u}_{j-1}) \cdot \vec{u}_j + (\alpha_j \vec{u}_j) \cdot \vec{u}_j + ... +$

$$+ (\alpha_k \vec{u}_k) \cdot \vec{u}_j$$

$\Rightarrow \vec{w} \cdot \vec{u}_j = \alpha_1 (\vec{u}_1 \vec{u}_j) + ... + \alpha_{j-1} (\vec{u}_{j-1} \cdot \vec{u}_j) + \alpha_j (\vec{u}_j \cdot \vec{u}_j) + ... +$

$$+ \alpha_k (\vec{u}_k \cdot \vec{u}_j)$$

$$\Rightarrow \quad \vec{w} \cdot \vec{u}_j = \alpha_j (\vec{u}_j \cdot \vec{u}_j)^1 \qquad\qquad \because \vec{u}_i \cdot \vec{u}_j = 0 \text{ for } i \neq j$$

$$\Rightarrow \quad \alpha_j = \frac{\vec{w} \cdot \vec{u}_j}{\vec{u}_j \cdot \vec{u}_j} \qquad\qquad \because \vec{u}_j \cdot \vec{u}_j > 0 \quad j = 1, 2, \ldots, k.$$

Remark : In the above theorem, If S is an orthonormal basis then $\vec{u}_j \cdot \vec{u}_j = 1$, for each $j = 1, 2, \ldots, k$, hence

$$\alpha_j = \vec{w} \cdot \vec{u}_j, \quad j = 1, 2, \ldots, k \qquad\blacksquare$$

Example 3.14 : Let $S = \{\vec{v}_1, \vec{v}_2, \vec{v}_3, \vec{v}_4\}$ be a set in $\mathbb{R}^4$, where,

$\vec{v}_1 = (1, -1, 2, -1)$, $\vec{v}_2 = (-2, 2, 3, 2)$, $\vec{v}_3 = (1, 2, 0, -1)$, $\vec{v}_4 = (1, 0, 0, 1)$

are an orthogonal basis for $\mathbb{R}^4$. Express $\vec{u} = (1, 1, 1, 1)$ as linear combination of vectors in S.

Solution : As S is orthogonal basis by theorem (7), we have,

$$\vec{u} = \alpha_1 \vec{v}_1 + \alpha_2 \vec{v}_2 + \alpha_3 \vec{v}_3 + \alpha_4 \vec{v}_4 \quad \text{where,}$$

$$\alpha_j = \frac{\vec{u} \cdot \vec{v}_j}{\vec{v}_j \cdot \vec{v}_j}, \quad j = 1, 2, 3, 4$$

$$\therefore \quad \alpha_1 = \frac{\vec{u} \cdot \vec{v}_1}{\vec{v}_1 \cdot \vec{v}_1} = \frac{(1, 1, 1, 1) \cdot (1, -1, 2, -1)}{(1, -1, 2, -1) \cdot (1, -1, 2, -1)} = \frac{1}{7}$$

$$\alpha_2 = \frac{\vec{u} \cdot \vec{v}_2}{\vec{v}_2 \cdot \vec{v}_2} = \frac{(1, 1, 1, 1) \cdot (-2, 2, 3, 2)}{(-2, 2, 3, 2) \cdot (-2, 2, 3, 2)} = \frac{5}{21}$$

$$\alpha_3 = \frac{\vec{u} \cdot \vec{v}_3}{\vec{v}_3 \cdot \vec{v}_3} = \frac{(1, 1, 1, 1) \cdot (1, 2, 0, -1)}{(1, 2, 0, -1) \cdot (1, 2, 0, -1)} = \frac{2}{6} = \frac{1}{3}$$

$$\alpha_4 = \frac{\vec{u} \cdot \vec{v}_4}{\vec{v}_4 \cdot \vec{v}_4} = \frac{(1, 1, 1, 1) \cdot (1, 0, 0, 1)}{(1, 0, 0, 1) \cdot (1, 0, 0, 1)} = \frac{2}{2} = 1$$

$$\therefore \vec{u} = \frac{1}{7}(1, -1, 2, -1) + \frac{5}{21}(-2, 2, 3, 2) + \frac{1}{3}(1, 2, 0, -1) + 1(1, 0, 0, 1)$$

Example 3.15 : Show that if W is subspace of $\mathbb{R}^n$ then

$$W \cap W^{\perp} = \{\vec{0}\}.$$

Solution : If $\vec{w} \in W \cap W^{\perp} \Rightarrow \vec{w} \in W$ and $\vec{w} \in W^{\perp}$

$\Rightarrow$ $\vec{w}$ is orthogonal to each vector in W in particular to itself

$\Rightarrow$ $\vec{w} \cdot \vec{w} = 0 \Leftrightarrow \vec{w} = 0$

Hence, $W \cap W^{\perp} = \{0\}$

3.3 Orthogonal Projections

We start this section with a simple problem which helps to define the title. Let $\vec{u} \in \mathbb{R}^n$ be a non-zero vector and $\vec{y} \in \mathbb{R}^n$ be any vector, then $\vec{y}$ can be written as sum of two vectors, one scalar multiple of $\vec{u}$ and the other orthogonal to $\vec{u}$.

Suppose $\vec{y} = \vec{x} + \vec{z}$ ... (i)

where, $\vec{x} = \alpha\vec{u}$ for some scalar and $\vec{z}$ is some vector orthogonal to $\vec{u}$. For any scalar α, let $\vec{z} = \vec{y} - \alpha\vec{u}$, so that (i) is satisfied. Then $\vec{y} - \alpha\vec{u}$ is orthogonal to $\vec{u}$ if and only if

$$0 = (y - \alpha\vec{u}) \cdot \vec{u} = \vec{y} \cdot \vec{u} - (\alpha\vec{u}) \cdot \vec{u}$$

$$= \vec{y} \cdot \vec{u} - \alpha(\vec{u} \cdot \vec{u})$$

Thus, equation (i) is satisfied with $\vec{z}$ orthogonal to $\vec{u}$ if and only if

$$\alpha = \frac{\vec{y} \cdot \vec{u}}{\vec{u} \cdot \vec{u}} \quad \text{and} \quad \vec{x} = \frac{\vec{y} \cdot \vec{u}}{\vec{u} \cdot \vec{u}}\,\vec{u}.$$

The vector $\vec{x} = \dfrac{\vec{y} \cdot \vec{u}}{\vec{u} \cdot \vec{u}}\,\vec{u}$ is usually denoted by $\hat{y}$ and is called the **orthogonal projection of $\vec{y}$ onto $\vec{u}$,** and the vector $\vec{z}$ is called the **component of $\vec{y}$ orthogonal to $\vec{u}$.** See the following figure.

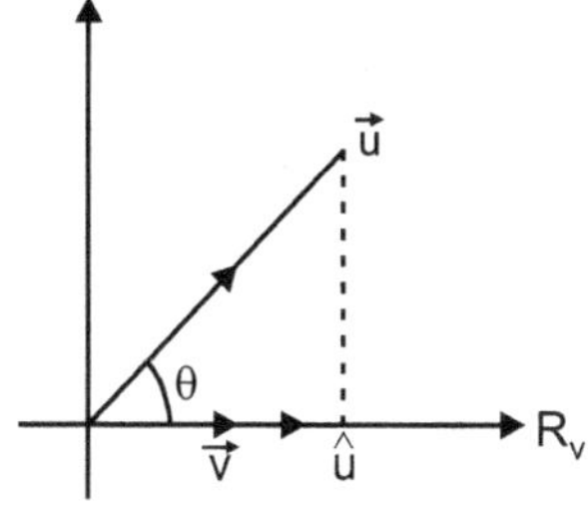

Fig. 3.7

Finding α to make $\vec{y} - \hat{y}$ orthogonal to $\vec{u}$.

Note : If δ is any non-zero scalar and if $\vec{u}$ is replaced by $\delta\vec{u}$ in the definition of $\hat{y}$, the orthogonal projection of $\vec{y}$ onto $\delta\vec{u}$ is exactly the same as the orthogonal projection of $\vec{y}$ onto $\vec{u}$. Hence, the projection is determined by the subspace L spanned by $\vec{u}$ (the line through $\vec{u}$ and $\vec{0}$). So we define

Orthogonal Projection :

Definition : If $\vec{v}$ is any non-zero vector in $\mathbb{R}^n$ and $\vec{u}$ be any vector in $\mathbb{R}^n$. Then the projection of $\vec{u}$ onto the line L spanned by $\vec{v}$, denoted by $\hat{u}$ or $\text{proj}_L \vec{u}$ and is defined by

$$\hat{u} = \text{proj}_L \vec{u} = \frac{\vec{u} \cdot \vec{v}}{\vec{v} \cdot \vec{v}} \vec{v}.$$

Example 3.16 : Let $\vec{u} = (3, 1, -7)$ and $\vec{v} = (1, 0, 5)$. Find the orthogonal projection of $\vec{u}$ along $\vec{v}$.

Solution : $\hat{u} = \text{Proj}_L(\vec{u}) = \frac{\vec{u} \cdot \vec{v}}{\vec{v} \cdot \vec{v}} \vec{v}$

where, L is line spanned by $\vec{v}$.

$$\Rightarrow \quad \hat{u} = \frac{(3, 1, -7) \cdot (1, 0, 5)}{(1, 0, 5) \cdot (1, 0, 5)} \vec{v} = \frac{3 + 0 - 35}{1 + 0 + 25} \vec{v}$$

$$\Rightarrow \quad \hat{u} = \frac{-32}{26} \vec{v} = \frac{-16}{13} (1, 0, 5)$$

$$\Rightarrow \quad \hat{u} = \left(\frac{-16}{13}, 0, \frac{-80}{13} \right)$$

Example 3.17 : If $\vec{u} = (1, 2, 3)$ and

$$W = \{\vec{v}_1 = (1, 0, 1), \vec{v}_2 = (-1, 0, 1)\}.$$

Find orthogonal projection of $\vec{u}$ onto W.

Solution : By definition of orthogonal projection

$$\hat{u} = \text{Proj}_W \vec{u} = \frac{\vec{u} \cdot \vec{v}_1}{\vec{v}_1 \cdot \vec{v}_1} \vec{v}_1 + \frac{\vec{u} \cdot \vec{v}_2}{\vec{v}_2 \cdot \vec{v}_2} \vec{v}_2$$

$$= \frac{(1, 2, 3) \cdot (1, 0, 1)}{(1, 0, 1) \cdot (1, 0, 1)} \vec{v}_1 + \frac{(1, 2, 3) \cdot (-1, 0, 1)}{(-1, 0, 1) \cdot (-1, 0, 1)} \vec{v}_2$$

$$= \frac{(1 + 0 + 3)}{1 + 0 + 1} \vec{v}_1 + \frac{(-1 + 0 + 3)}{(-1)(-1) + 0 + 1} \vec{v}_2$$

$$= \frac{4}{2} (1, 0, 1) + \frac{2}{2} (-1, 0, 1)$$

$$= (2, 0, 2) + (-1, 0, 1) = (1, 0, 3)$$

$$\therefore \quad \hat{u} = \text{Proj}_W \vec{u} = (1, 0, 3)$$

Theorem 8 If U is $m \times n$ matrix then it has orthonormal columns if and only if $U^t U = I_3$.

Proof : Without loss of generality, let U has 3 columns each vector in $\mathbb{R}^m$.

Now, $\qquad U = [\vec{u}_1 \ \vec{u}_2 \ \vec{u}_3]$

$$\therefore \qquad U^t U = \begin{bmatrix} \vec{u}_1^t \\ \vec{u}_2^t \\ \vec{u}_3^t \end{bmatrix}_{3 \times 1} \underset{\underset{U}{\uparrow}}{[\vec{u}_1 \ \vec{u}_2 \ \vec{u}_3]_{1 \times 3}}$$

$$\underset{\underset{U^t}{\uparrow}}{}$$

$$= \begin{bmatrix} \vec{u}_1^t \vec{u}_1 & \vec{u}_1^t \vec{u}_2 & \vec{u}_1^t \vec{u}_3 \\ \vec{u}_2^t \vec{u}_1 & \vec{u}_2^t \vec{u}_2 & \vec{u}_2^t \vec{u}_3 \\ \vec{u}_3^t \vec{u}_1 & \vec{u}_3^t \vec{u}_2 & \vec{u}_3^t \vec{u}_3 \end{bmatrix}$$

Now, column of U are orthonormal if and only if $\vec{u}_i^t \vec{u}_j = 0$, $i \neq j$ and $\vec{u}_i^t \cdot \vec{u}_j = 1$, $i = j$.

So that the matrix $U^t U$ becomes identity matrix of order 3×3, hence $U^t U = I_3$. ∎

Theorem 9 Let U be an $m \times n$ matrix with orthonormal columns and let $X, Y \in \mathbb{R}^n$ be any vectors. Then

    (a) $(UX) \cdot (UY) = X \cdot Y$

    (b) $\|UX\| = \|X\|$

    (c) $(UX) \cdot (UY) = 0 \Leftrightarrow X \cdot Y = 0$

Proof : Given U be an $m \times n$ matrix with orthonormal columns and $X, Y \in \mathbb{R}^n$ be arbitrary vectors :

(a) To show that, $(UX) \cdot (UY) = X \cdot Y$

Now, $U = \{\vec{u}_1 \ \vec{u}_2 \ \dots \ \vec{u}_n\}$ with $\vec{u}_i \cdot \vec{u}_j = \begin{cases} 0 & i \neq j \\ 1 & i = j \end{cases}$

$$UX = [\vec{u}_1 \ \vec{u}_2 \ \dots \ \vec{u}_n]_{m \times n} \begin{bmatrix} x_1 \\ x_2 \\ \vdots \\ x_n \end{bmatrix}_{n \times 1} \qquad UY = [\vec{u}_1 \ \vec{u}_2 \ \dots \ \vec{u}_n] \begin{bmatrix} y_1 \\ y_2 \\ \vdots \\ y_n \end{bmatrix}_{n \times 1}$$

$\therefore \quad (UX) \cdot (UY) = (\vec{u}_1 x_1) \cdot (\vec{u}_1 y_1) + (\vec{u}_2 x_2) \cdot (\vec{u}_2 y_2) + \dots + (\vec{u}_n x_n) \cdot (\vec{u}_n y_n)$

$\Rightarrow \quad (UX) \cdot (UY) = (\vec{u}_1 \cdot \vec{u}_1)(x_1 y_1) + (\vec{u}_2 \cdot \vec{u}_2) x_2 y_2 + \dots +$

$$\dots (\vec{u}_n \cdot \vec{u}_n) x_n y_n$$

$\Rightarrow \quad (UX) \cdot (UY) = x_1 y_1 + x_2 y_2 + \dots + x_n y_n = X \cdot Y \quad \because \vec{u}_i \cdot \vec{u}_i = 1, \ 1 \leq i \leq n$

$\therefore \quad (UX) \cdot (UY) = X \cdot Y$

(b) To show that, $\|UX\| = \|x\|$

Put $Y = X$ in (a) we have,

$$(UX) \cdot (UY) = X \cdot X$$

$\Rightarrow \qquad\qquad \|UX\|^2 = \|X\|^2 \qquad\qquad\qquad \vec{u} \cdot \vec{u} = \|\vec{u}\|^2$

$\Rightarrow \qquad\qquad \|UX\| = \|X\|$

(c) From (a) $(UX) \cdot (UY) = X \cdot Y$

$\therefore \qquad (UX) \cdot (UY) = 0 \Leftrightarrow X \cdot Y = 0$ ■

Example 3.18 : Let $U = \begin{bmatrix} \dfrac{1}{\sqrt{2}} & \dfrac{-1}{\sqrt{2}} \\ 0 & 0 \\ \dfrac{1}{\sqrt{2}} & \dfrac{1}{\sqrt{2}} \end{bmatrix}$ and $X = \begin{bmatrix} 1 \\ 0 \end{bmatrix}$, verify that U

has orthonormal columns and $\|UX\| = \|x\|$.

Solution : Given : $U = \begin{bmatrix} \dfrac{1}{\sqrt{2}} & \dfrac{-1}{\sqrt{2}} \\ 0 & 0 \\ \dfrac{1}{\sqrt{2}} & \dfrac{1}{\sqrt{2}} \end{bmatrix}$

$$\underset{\overset{\uparrow}{\vec{u}_1}}{} \quad \underset{\overset{\uparrow}{\vec{u}_2}}{}$$

Now, $\vec{u}_1 \cdot \vec{u}_2 = \left(\dfrac{1}{\sqrt{2}}\right)\left(\dfrac{-1}{\sqrt{2}}\right) + 0 + \left(\dfrac{1}{\sqrt{2}}\right)\left(\dfrac{1}{\sqrt{2}}\right) = \dfrac{-1}{2} + 0 + \dfrac{1}{2} = 0$

$\therefore$ $\vec{u}_1$ and $\vec{u}_2$ are orthogonal with

$$\|\vec{u}_1\| = \sqrt{\left(\dfrac{1}{\sqrt{2}}\right)^2 + 0 + \left(\dfrac{1}{\sqrt{2}}\right)^2} = \sqrt{\dfrac{1}{2} + 0 + \dfrac{1}{2}} = 1$$

and $$\|\vec{u}_2\| = \sqrt{\left(\dfrac{-1}{\sqrt{2}}\right)^2 + 0 + \left(\dfrac{1}{\sqrt{2}}\right)^2} = \sqrt{\dfrac{1}{2} + 0 + \dfrac{1}{2}} = 1$$

Hence, U has orthonormal columns.

Now, $\quad UX = \underset{\overset{\uparrow}{\vec{U}}}{\begin{bmatrix} \dfrac{1}{\sqrt{2}} & \dfrac{-1}{\sqrt{2}} \\ 0 & 0 \\ \dfrac{1}{\sqrt{2}} & \dfrac{1}{\sqrt{2}} \end{bmatrix}_{3\times 2}} \underset{\overset{\uparrow}{X}}{\begin{bmatrix} 1 \\ 0 \end{bmatrix}_{2\times 1}} = \begin{bmatrix} \dfrac{1}{\sqrt{2}} & 0 & \dfrac{1}{\sqrt{2}} \end{bmatrix}$

$\therefore$ $\quad \|UX\| = \sqrt{\left(\dfrac{1}{\sqrt{2}}\right)^2 + 0 + \left(\dfrac{1}{\sqrt{2}}\right)^2} = 1$ and $\|X\| = \sqrt{1+0} = 1$

Hence, $\quad \|UX\| = \|x\|$

Example 3.19 : *Determine whether the vectors $\vec{u}_1,\ \vec{u}_2,\ \vec{u}_3$ are orthogonal in $\mathbb{R}^3$, where*

$$\vec{u}_1 = \begin{bmatrix} \dfrac{1}{\sqrt{18}} \\ \dfrac{4}{\sqrt{18}} \\ \dfrac{1}{\sqrt{18}} \end{bmatrix}, \quad \vec{u}_2 = \begin{bmatrix} \dfrac{1}{\sqrt{2}} \\ 0 \\ \dfrac{-1}{\sqrt{2}} \end{bmatrix}, \quad \vec{u}_3 = \begin{bmatrix} \dfrac{-2}{3} \\ \dfrac{1}{3} \\ \dfrac{-2}{3} \end{bmatrix}$$

Solution : We have,

$$\vec{u}_1 \cdot \vec{u}_2 = \frac{1}{\sqrt{18}} \cdot \frac{1}{\sqrt{2}} + 0 - \frac{1}{\sqrt{18}} \cdot \frac{1}{\sqrt{2}} = 0$$

$$\vec{u}_2 \cdot \vec{u}_3 = \frac{1}{\sqrt{2}} \cdot \frac{-2}{3} + 0 + \frac{-1}{\sqrt{2}} \cdot \frac{-2}{3} = 0$$

and $\quad \vec{u}_1 \cdot \vec{u}_3 = \dfrac{1}{\sqrt{18}} \cdot \dfrac{-2}{3} + \dfrac{4}{\sqrt{18}} \cdot \dfrac{1}{3} + \dfrac{1}{\sqrt{18}} \cdot \dfrac{-2}{3} = 0$

This shows that $\vec{u}_1$, $\vec{u}_2$ and $\vec{u}_3$ are orthogonal vectors; next we check whether the norm of each of these vectors is 1. For this, we have

$$\|\vec{u}_1\| = \sqrt{\left(\frac{1}{\sqrt{18}}\right)^2 + \left(\frac{4}{\sqrt{18}}\right)^2 + \left(\frac{1}{\sqrt{18}}\right)^2} = \sqrt{\frac{1}{18} + \frac{16}{18} + \frac{1}{18}} = 1$$

$$\|\vec{u}_2\| = \sqrt{\left(\frac{1}{\sqrt{2}}\right)^2 + 0 + \left(\frac{-1}{\sqrt{2}}\right)^2} = \sqrt{\frac{1}{2} + \frac{1}{2}} = 1$$

and $\quad \|\vec{u}_3\| = \sqrt{\left(\frac{-2}{3}\right)^2 + \left(\frac{1}{3}\right)^2 + \left(\frac{-2}{3}\right)^2} = \sqrt{\frac{4}{9} + \frac{1}{9} + \frac{4}{9}} = 1$

Thus, we see that the vectors $\vec{u}_1$, $\vec{u}_2$, $\vec{u}_3$ are orthogonal.

Orthogonal Projections Continued :

We have seen the orthogonal projection of a vector $\vec{u}$ in $\mathbb{R}^n$ onto the line passing through $\vec{u}$ and $\vec{0}$. The line is a subspace of $\mathbb{R}^n$ spanned by $\vec{u}$. We generalize this fact.

Given a vector $\vec{u}$ and its orthogonal projection onto a subspace W (may not be a line).

First we have two examples, which helps us to define the orthogonal projection of a vector onto a subspace.

Example 3.20 : Let $\{\vec{u}_1, \vec{u}_2, ..., \vec{u}_n\}$ be a basis for $\mathbb{R}^n$ and $\vec{v}$ be any vector in $\mathbb{R}^n$, then $\vec{v}$ can be expressed as sum of two vectors.

Solution : Since $\{\vec{u}_1, \vec{u}_2, ..., \vec{u}_n\}$ is a basis for $\mathbb{R}^n$, $\vec{v}$ can be written as linear combination of $\vec{u}_1, \vec{u}_2, ..., \vec{u}_n$, say

$$\vec{v} = \alpha_1 \vec{u}_1 + \alpha_2 \vec{u}_2 + ... + \alpha_n \vec{u}_n$$

we group the right hand sum in two parts as

$$\vec{v} = (\alpha_1\vec{u}_1 + \ldots + \alpha_i\vec{u}_i) + (\alpha_{i+1}\vec{u}_{i+1} + \ldots + \alpha_n\vec{u}_n)$$

$$= \vec{z}_1 + \vec{z}_2$$

where, $\vec{z}_1 = \alpha_1\vec{u}_1 + \ldots + \alpha_i\vec{u}_i$ and

$$\vec{z}_2 = \alpha_{i+1}\vec{u}_{i+1} + \ldots + \alpha_n\vec{u}_n$$

If the basis of $\mathbb{R}^n$ is orthogonal, then above expression for $\vec{v}$ as sum of two vectors is significant. See the following example.

Example 3.21 : Let $\{\vec{u}_1, \vec{u}_2, \ldots, \vec{u}_7\}$ be the orthogonal basis for $\mathbb{R}^7$ and W be a subspace of $\mathbb{R}^7$ spanned by $\vec{u}_1, \vec{u}_2, \vec{u}_3$. Then any vector $\vec{u}$ in $\mathbb{R}^n$ can be written as $\vec{u} = \vec{z}_1 + \vec{z}_2$, where $\vec{z}_1$ is in W and $\vec{z}_2$ is in $W^\perp$.

Solution : As $\{\vec{u}_1, \vec{u}_2, \ldots, \vec{u}_7\}$ is an orthogonal basis for $\mathbb{R}^7$, for any $\vec{u}$ in $\mathbb{R}^7$, we have

$$\vec{u} = \alpha_1\vec{u}_1 + \alpha_2\vec{u}_2 + \ldots + \alpha_7\vec{u}_7$$

where, $\alpha_1, \alpha_2, \ldots, \alpha_7$ are scalars.

We can write $\vec{u}$ as,

$$\vec{u} = (\alpha_1\vec{u}_1 + \alpha_2\vec{u}_2 + \alpha_3\vec{u}_3) + (\alpha_4\vec{u}_4 + \ldots + \alpha_7\vec{u}_7)$$

$$= \vec{z}_1 + \vec{z}_2$$

where, $\vec{z}_1 = \alpha_1\vec{u}_1 + \alpha_2\vec{u}_2 + \alpha_3\vec{u}_3$ and

$$\vec{z}_2 = \alpha_4\vec{u}_4 + \ldots + \alpha_7\vec{u}_7$$

Clearly, $\vec{z}_1$ is in W; to show that $\vec{z}_2$ is in $W^\perp$, we have to show that $\vec{z}_2$ is orthogonal to each basis vector of W. We have,

$$\vec{z}_2 \cdot \vec{u}_1 = (\alpha_2\vec{u}_4 + \ldots + \alpha_7\vec{u}_7) \cdot \vec{u}_1$$

$$= \alpha_4(\vec{u}_4 \cdot \vec{u}_1) + \ldots + \alpha_7(\vec{u}_7 \cdot \vec{u}_1)$$

$$= \alpha_4 \cdot 0 + \ldots + \alpha_7 \cdot 0 = 0$$

Since, the given basis is orthogonal $\vec{u}_i \cdot \vec{u}_j = 0$ for all $i \neq j$.

Similarly, we can show that $\vec{z}_2 \cdot \vec{u}_2 = 0$, $\vec{z}_2 \cdot \vec{u}_3 = 0$

Thus, $\vec{u} = \vec{z}_1 + \vec{z}_2$ with $\vec{z}_1$ in W and $\vec{z}_2$ in $W^\perp$.

Now, we state and prove the **Orthogonal Decomposition Theorem,** which provides the generalization of orthogonal projection.

Theorem 10 Orthogonal Decomposition Theorem

If W is any subspace of $\mathbb{R}^n$. Then each $\vec{w}$ in $\mathbb{R}^n$ can be written uniquely in the form $\vec{w} = \hat{w} + \vec{z}$, where $\hat{w} \in W$ and $\vec{z} \in W^{\perp}$.

Moreover, if $B = \{\vec{u}_1, \vec{u}_2, \ldots, \vec{u}_k\}$ is any orthogonal basis of W then,

$$\hat{w} = \frac{\vec{w} \cdot \vec{u}_1}{\vec{u}_1 \cdot \vec{u}_1}\vec{u}_1 + \frac{\vec{w} \cdot \vec{u}_2}{\vec{u}_2 \cdot \vec{u}_2}\vec{u}_2 + \ldots + \frac{\vec{w} \cdot \vec{u}_k}{\vec{u}_k \cdot \vec{u}_k}\vec{u}_k \text{ and } \vec{z} = \vec{w} - \hat{w}$$

Proof : Step I : Let $B = \{\vec{u}_1, \vec{u}_2, \ldots, \vec{u}_k\}$ be any orthogonal basis for W then by definition of orthogonal Projection.

$$\hat{w} = \frac{\vec{w} \cdot \vec{u}_1}{\vec{u}_1 \cdot \vec{u}_1}\vec{u}_1 + \frac{\vec{w} \cdot \vec{u}_2}{\vec{u}_2 \cdot \vec{u}_2}\vec{u}_2 + \ldots + \frac{\vec{w} \cdot \vec{u}_k}{\vec{u}_k \cdot \vec{u}_k}\vec{u}_k$$

Define, $\vec{z} = \vec{w} - \hat{w}$

Then $\vec{z} \cdot \vec{u}_1$, gives,

$$\vec{z} \cdot \vec{u}_1 = (\vec{w} - \hat{w}) \cdot \vec{u}_1 = \vec{w} \cdot \vec{u}_1 - \left[\frac{\vec{w} \cdot \vec{u}_1}{\vec{u}_1 \cdot \vec{u}_1}\vec{u}_1 + \ldots + \frac{\vec{w} \cdot \vec{u}_k}{\vec{u}_k \cdot \vec{u}_k}\vec{u}_k\right] \cdot \vec{u}_1$$

$$\Rightarrow \vec{z} \cdot \vec{u}_1 = \vec{w} \cdot \vec{u}_1 - \left(\frac{\vec{w} \cdot \vec{u}_1}{\vec{u}_1 \cdot \vec{u}_1}\right)\vec{u}_1 \cdot \vec{u}_1 + 0 + \ldots + 0 \qquad \ldots B \text{ is orthogonal}$$

$$\Rightarrow \vec{z} \cdot \vec{u}_1 = \vec{w} \cdot \vec{u}_1 - \vec{w} \cdot \vec{u}_1 = 0$$

Similarly, $\vec{z} \cdot \vec{u}_2 = 0 = \vec{z} \cdot \vec{u}_3 = \ldots = \vec{z} \cdot \vec{u}_k$

i.e. $\vec{z}$ is orthogonal to every vector in B and as B is basis for W,

$\Rightarrow \vec{z}$ is orthogonal to every vector in $W \Rightarrow \vec{z} \in W^{\perp}$

$\therefore \quad \vec{w} = \hat{w} + \vec{z}$ where, $\hat{w} \in W$ and $\vec{z} \in W^{\perp}$.

Step II : To show that, uniqueness. Suppose $\vec{w} = \hat{w}_1 + \vec{z}_1$ where

$\hat{w}_1 \in W$ and $\vec{z}_1 \in W^{\perp}$

$\Rightarrow \quad \hat{w}_1 + \vec{z}_1 = \hat{w} + \vec{z} \Rightarrow \hat{w} - \hat{w}_1 = \vec{z}_1 - \vec{z} \ \varepsilon \ W \cap W^{\perp}$

But $W \cap W^{\perp} = \{0\} \Rightarrow \hat{w} - \hat{w}_1 = 0 = \vec{z}_1 - \vec{z}$ i.e. $\hat{w} = \hat{w}_1$ and $\vec{z}_1 = \vec{z}$

$\therefore \quad$ The representation is unique. ∎

Definition : Let W be a subspace of $\mathbb{R}^n$, and $\{\vec{u}_1, \vec{u}_2, ..., \vec{u}_k\}$ be an orthogonal basis for W; then any vector $\vec{u}$ in $\mathbb{R}^n$ is written uniquely as

$$\vec{u} = \hat{u} + \vec{z}, \text{ where } \vec{z} \in W^{\perp} \text{ and}$$

$$\hat{u} = \left(\frac{\vec{u} \cdot \vec{u}_1}{\vec{u}_1 \cdot \vec{u}_1}\right)\vec{u}_1 + \left(\frac{\vec{u} \cdot \vec{u}_2}{\vec{u}_2 \cdot \vec{u}_2}\right)\vec{u}_2 + ... + \left(\frac{\vec{u} \cdot \vec{u}_k}{\vec{u}_k \cdot \vec{u}_k}\right)\vec{u}_k \qquad ... \text{(i)}$$

in W. The vector $\hat{u}$ in W as given in equation (i) is called the **orthogonal projection of $\vec{u}$ onto W** and is written as

$$\hat{u} = \text{proj}_W \vec{u}$$

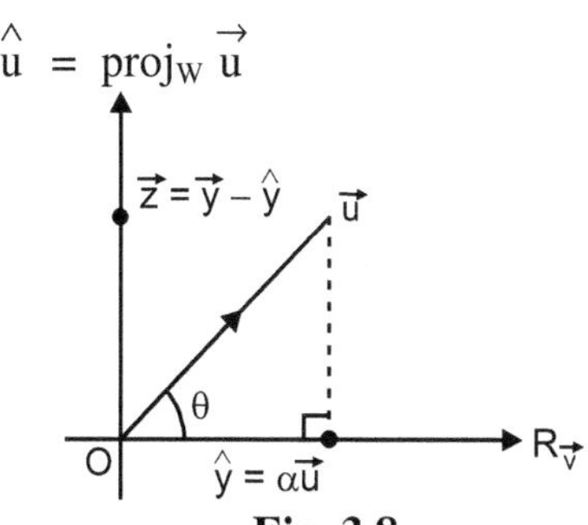

Fig. 3.8

Example 3.22 : *Let $S = \{\vec{u}_1, \vec{u}_2, \vec{u}_3\}$ be an orthogonal basis for $\mathbb{R}^3$,*

where, $\vec{u}_1 = \begin{bmatrix} 0 \\ 1 \\ 0 \end{bmatrix}, \vec{u}_2 = \begin{bmatrix} \frac{1}{\sqrt{2}} \\ 0 \\ \frac{1}{\sqrt{2}} \end{bmatrix}, \vec{u}_3 = \begin{bmatrix} \frac{1}{\sqrt{2}} \\ 0 \\ \frac{-1}{\sqrt{2}} \end{bmatrix}.$ *Write $\vec{y} = (1, 2, 3)^t$ as the*

sum of vectors in S and a vector orthogonal to S.

Solution : The orthogonal projection of $\vec{y}$ onto S is

$$\hat{y} = \frac{\vec{y} \cdot \vec{u}_1}{\vec{u}_1 \cdot \vec{u}_1}\vec{u}_1 + \frac{\vec{y} \cdot \vec{u}_2}{\vec{u}_2 \cdot \vec{u}_2}\vec{u}_2 + \frac{\vec{y} \cdot \vec{u}_3}{\vec{u}_3 \cdot \vec{u}_3}\vec{u}_3$$

$$\Rightarrow \hat{y} = \frac{(1, 2, 3) \cdot (0, 1, 0)}{1}\vec{u}_1 + \frac{(1, 2, 3) \cdot \left(\frac{1}{\sqrt{2}}, 0, \frac{1}{\sqrt{2}}\right)}{1}\vec{u}_2$$

$$+ \frac{(1, 2, 3) \cdot \left(\frac{1}{\sqrt{2}}, 0, \frac{-1}{\sqrt{2}}\right)}{1}\vec{u}_3$$

$$\Rightarrow \quad \hat{y} = 2\vec{u}_1 + (2\sqrt{2})\,\vec{u}_2 + (-\sqrt{2})\,\vec{u}_3$$

$$\Rightarrow \quad \hat{y} = 2(0,\,1,\,0) + 2\sqrt{2}\left(\frac{1}{\sqrt{2}},\,0,\,\frac{1}{\sqrt{2}}\right) - \sqrt{2}\left(\frac{1}{\sqrt{2}},\,0,\,\frac{-1}{\sqrt{2}}\right)$$

$$\hat{y} = (0,\,2,\,0) + (2,\,0,\,2) + (-1,\,0,\,1)$$

$$\therefore \quad \hat{y} = (1,\,2,\,3)$$

$$\Rightarrow \quad \vec{z} = y - \hat{y} = (1,\,2,\,3) - (1,\,2,\,3) = (0,\,0,\,0) \text{ which is orthogonal}$$

to S.

Note : From above example, we see that the vector $\vec{y}$ is in $\mathbb{R}^3$, then $\text{proj}_W\,\vec{y} = \vec{y}$ itself.

$\boxed{\textbf{Theorem 11}}$ If $\{\vec{u}_1,\,\vec{u}_2,\,\ldots,\,\vec{u}_k\}$ is an orthonormal basis for a subspace W of $\mathbb{R}^n$ then $\text{Proj}_W\,\vec{u} = (\vec{u}\cdot\vec{u}_1)\,\vec{u}_1 + (\vec{u}\cdot\vec{u}_2)\,\vec{u}_2 + \ldots + (\vec{u}\cdot\vec{u}_k)\,\vec{u}_k$ moreover if, $U = [\vec{u}_1\;\vec{u}_2\;\ldots\;\vec{u}_k]$ then $\text{Proj}_W\,\vec{u} = UU^t\,\vec{u}$, $\forall\,\vec{u}\in\mathbb{R}^n$.

Proof : Given that, $B = \{\vec{u}_1,\,\vec{u}_2,\,\ldots,\,\vec{u}_k\}$ is an orthonormal basis for a subspace W of $\mathbb{R}^n \Rightarrow \vec{u}_i\cdot\vec{u}_j = \begin{cases} 0 & i\neq j \\ 1 & i=j \end{cases}$

$\therefore$ By Orthogonal Decomposition Theorem,

$$\text{Proj}_W\,\vec{u} = (\vec{u}\cdot\vec{u}_1)\,\vec{u}_1 + (\vec{u}\cdot\vec{u}_2)\,\vec{u}_2 + \ldots + (\vec{u}\cdot\vec{u}_k)\,\vec{u}_k \qquad \ldots \text{(I)}$$

$$\text{Now, } UU^t\,\vec{u} = [\vec{u}_1\;\vec{u}_2\;\vec{u}_3\;\ldots\;\vec{u}_k] \begin{bmatrix} \vec{u}_1^{\,t} \\ \vec{u}_2^{\,t} \\ \vdots \\ \vec{u}_k^{\,t} \end{bmatrix} ,\; \vec{u}\in\mathbb{R}^n$$

$$\underset{U_{n\times k}\text{ matrix}}{\uparrow} \qquad \underset{k\times n\text{ matrix}}{}$$

$$\Rightarrow \quad UU^t\,\vec{u} = [\vec{u}_1\;\vec{u}_2\;\ldots\;\vec{u}_k]_{n\times k} \begin{bmatrix} \vec{u}_1^{\,t}\,\vec{u} \\ \vec{u}_2^{\,t}\,\vec{u} \\ \vdots \\ \vec{u}_k^{\,t}\,\vec{u} \end{bmatrix}$$

By property of orthogonally,

$$UU^t\,\vec{u} = [\vec{u}_1^{\,t}\cdot\vec{u}]\,\vec{u}_1 + [\vec{u}_2^{\,t}\cdot\vec{u}]\,\vec{u}_2 + \ldots + [\vec{u}_k^{\,t}\cdot\vec{u}]\,\vec{u}_k$$

By property of commutativity of dot product we have,

$$UU^t\,\vec{u} = (\vec{u} \cdot \vec{u}_1)\,\vec{u}_1 + (\vec{u} \cdot \vec{u}_2)\,\vec{u}_2 + \ldots + (\vec{u} \cdot \vec{u}_k)\,\vec{u}_k \qquad \ldots \text{(II)}$$

From equations (I) and (II),

$$\text{Proj}_W\,\vec{u} = UU^t\,\vec{u} \qquad\blacksquare$$

Example 3.23 : *Let W be subspace of $\mathbb{R}^n$ and $\vec{v}, \vec{w}$ be vectors in $\mathbb{R}^n$ and $\vec{u} = \vec{v} + \vec{w}$. If $\vec{u}_1$ is the projection of $\vec{v}$ onto W and $\vec{v}_1$ is the Projection of $\vec{w}$ onto W then show that, $\vec{u}_1 + \vec{v}_1$ is the projection of $\vec{u}$ onto W.*

Solution : Given that W is subspace of $\mathbb{R}^n$. Let $\vec{u}$ be orthonormal matrix having columns as orthonormal basis for W.

Then by previous theorem,

$$\begin{aligned}
\text{Proj}_W\,\vec{u} &= UU^t\,\vec{u} \\[4pt]
&= UU^t\,(\vec{v} + \vec{w}) \qquad\qquad \ldots\ \vec{u} + \vec{w} \\[4pt]
&= UU^t\,\vec{v} + UU^t\,\vec{w}
\end{aligned}$$

$$\ldots \text{Matrix multiplication is distributive}$$

$$\begin{aligned}
&= \text{Proj}_W\,\vec{v} + \text{Proj}_W\,\vec{w} \\[4pt]
&= \vec{u}_1 + \vec{v}_1 \\[4pt]
\Rightarrow \qquad \vec{u}_1 + \vec{v}_1 &= \text{Proj}_W\,\vec{u}
\end{aligned}$$

3.4 Symmetric Matrices and their Diagonalization

We know that, if A is $m \times n$ matrix then its transpose A^T is $n \times m$ matrix which is obtained by the interchanging rows into columns and columns into rows.

For example : (1) If $A = \begin{bmatrix} 2 & 1 & 2 \\ 0 & 1 & 0 \end{bmatrix}_{2 \times 3} \Rightarrow A^t = \begin{bmatrix} 2 & 0 \\ 1 & 1 \\ 2 & 0 \end{bmatrix}_{3 \times 2}$

(2) If A is $1 \times n$ row vector in $\mathbb{R}^n$ then A^t is $n \times 1$ column vector in $\mathbb{R}^n$.

(3) $A = \begin{bmatrix} 1 & 1 & 2 \\ 1 & 2 & 4 \\ 2 & 4 & 3 \end{bmatrix}_{3 \times 3} \Rightarrow A^t = \begin{bmatrix} 1 & 1 & 2 \\ 1 & 2 & 4 \\ 2 & 4 & 3 \end{bmatrix}_{3 \times 3}$

In the above examples, if we see example (3), we have the property that $A = A^t$; such matrices are known as symmetric matrices.

Definition :

A square matrix A of order $n \times n$ is called symmetric matrix if $A^t = A$.

Note that, if A is symmetric matrix then A must be first of all a square matrix with entries occur in pairs on the opposite sides of the main diagonal i.e., if A is symmetric matrix and $A = [a_{ij}]_{n \times n}$ then $a_{ij} = a_{ji} \ \forall \ 1 \le i, j \le n$.

Example 3.24 : *Show that if A is symmetric then A^2 is also symmetric.*

Solution : Given that, A is symmetric matrix $\Rightarrow A^t = A$.

Now, consider
$$(A^2)^t = (AA)^t = A^t A^t \qquad \ldots (AB)^t = B^t A^t$$
$$= AA \qquad \ldots A^t = A$$
$$= A^2$$
$$\therefore \qquad (A^2)^t = A^2$$

Hence, A^2 is also symmetric.

Remark : Can we generalize the above statement that of A is symmetric matrix then for $k \ge 1$, A^k is also symmetric ?

Example 3.25 : *Suppose A is $n \times n$ symmetric matrix and B is any $n \times m$ matrix then $B^t AB$, $B^t B$ and BB^t are symmetric matrices.*

Solution : Given that A is $n \times n$ symmetric matrix $\Rightarrow A = A^t$.

Now, (i)
$$(B^t AB)^t = B^t A^t (B^t)^t$$

$\qquad\qquad\qquad\qquad\qquad\qquad$ ... Property of transpose of matrices

$$= B^t AB \qquad \ldots A^t = A \text{ and } (B^t)^t = B$$
$$\Rightarrow \qquad (B^t AB)^t = B^t AB \qquad\qquad \text{hence it is symmetric}$$
$$\text{(ii)} \qquad (B^t B)^t = B^t (B^t)^t \ \ldots \text{Property of symmetric matrices}$$
$$= B^t B$$
$$\therefore \qquad (B^t B)^t = B^t B \qquad\qquad \text{hence it is symmetric matrix}$$

Similarly, one show BB^t is also symmetric matrix.

Example 3.26 : *Show that if A is an $n \times n$ symmetric matrix, then $(AX) \cdot Y = X \cdot (AY)$, $\forall X, Y \in \mathbb{R}^n$.*

Solution : Given that, A is symmetric matrix $\Rightarrow A = A^t$

Now,
$$(AX) \cdot Y = (AX)^t Y \qquad \ldots \text{Definition of dot product}$$
$$= (X^t A^t) Y \qquad \ldots \text{Property of transpose}$$
$$= X^t (A^t Y) \qquad\qquad \ldots \text{By associativity}$$

$$= X^t (AY) \qquad \qquad \ldots A^t = A$$
$$= X \cdot (AY) \qquad \ldots \text{Definition of dot product}$$

Hence, $\qquad (AX) \cdot Y = X \cdot (AY)$

Example 3.27 : *Show that if A is n $\times$ n invertible symmetric matrix then A^{-1} is also symmetric.*

Solution : We know the property of symmetric matrix that $A = A^t$.

To show that, $\qquad A^{-1} = (A^{-1})^t$

$\because \quad$ A is invertible $\Rightarrow AA^{-1} = I_{n \times n}$

But $I_{n \times n}$ is symmetric $\Rightarrow I^t = I$

$$\therefore \qquad\qquad (AA^{-1}) = I = I^t = (AA^{-1})^t$$
$$\Rightarrow \qquad\qquad (AA^{-1}) = (AA^{-1})^t$$
$$= (A^{-1})^t \cdot A^t \qquad\qquad \ldots (AB)^t = B^t A^t$$
$$\therefore \qquad AA^{-1} = (A^{-1})^t A \qquad\qquad \ldots A^t = A$$

Also, $\qquad AA^{-1} = A^{-1}A = I_{n \times n}$

$$\therefore \qquad A^{-1}A = AA^{-1} = (A^{-1})^t A$$
$$\Rightarrow \qquad A^{-1}A = (A^{-1})^t A$$
$$\Rightarrow \qquad (A^{-1}A)\, A^{-1} = ((A^{-1})^t\, AA^{-1} \qquad \ldots \text{Post multiply by } A^{-1}$$
$$\Rightarrow \qquad A^{-1}(AA^{-1}) = (A^{-1})^t\, I \qquad \ldots AA^{-1} = I \text{ and associativity}$$
$$\Rightarrow \qquad A^{-1} I = (A^{-1})^t$$
$$\Rightarrow \qquad A^{-1} = (A^{-1})^t$$

Hence, A^{-1} is also symmetric matrix.

Note :

(1) If A is invertible n $\times$ n orthogonal matrix then

$AA^t = I_{n \times n} = A^t A \Rightarrow A^{-1} = A^t$ i.e. for any invertible square orthogonal matrix A, $A^t = A^{-1}$.

(2) As seen in the previous chapter, if A is diagonalizable matrix then there exist matrix P, such that $A = PDP^{-1}$ where D is diagonal matrix with diagonal entries are eigen values of A.

Definition : (Orthogonally Diagonalizable)

An n $\times$ n matrix A is said to be orthogonally diagonalizable if there is an orthogonal symmetric matrix $P(P^t = P^{-1})$ and a diagonal matrix D, such that $A = PDP^{-1} = PDP^t$.

Theorem 12 If A is symmetric matrix of order n $\times$ n then any two eigen vectors from different eigen space are orthogonal.

Proof : Let A be any $n \times n$ symmetric matrix and $\vec{v}_1, \vec{v}_2$ are two eigen vectors corresponding two distinct eigen valves (say) l_1, l_2 respectively.

To show that, Eigen space corresponding to $\vec{v}_1$ and $\vec{v}_2$ are orthogonal it is equivalent to show that, $\vec{v}_1 \cdot \vec{v}_2 = 0$ (WLOG, assume l_1, l_2 are non-zero).

Now consider,

$$
\begin{aligned}
l_1 \vec{v}_1 \cdot \vec{v}_2 &= (l_1\vec{v}_1)^t \, \vec{v}_2 &&\ldots \text{ By definition of dot product} \\
&= (A\vec{v}_1)^t \, \vec{v}_2 &&\ldots A\vec{v}_1 = l_1\vec{v}_1 \\
&= (\vec{v}_1^{\,t} A^t) \, \vec{v}_2 &&\ldots (AB)^t = B^tA^t \\
&= \vec{v}_1^{\,t} (A^t\vec{v}_2) &&\ldots \text{ Associativity} \\
&= v_1^{\,t} (A\vec{v}_2) &&\ldots A = A^t \\
&= \vec{v}_1^{\,t} (l_2\vec{v}_2) &&\ldots A\vec{v}_2 = l_2\vec{v}_2 \\
&= l_2 (\vec{v}_1^{\,t} \vec{v}_2) &&\ldots \text{ Property of scalar} \\
&= l_2 (\vec{v}_1 \cdot \vec{v}_2) &&\ldots \text{ By definition of dot product}
\end{aligned}
$$

$$\Rightarrow \qquad l_1 \vec{v}_1 \cdot \vec{v}_2 = l_2 \vec{v}_1 \cdot \vec{v}_2 \text{ with } l_1 \neq l_2$$

$$\Rightarrow \qquad (l_1 - l_2) \vec{v}_1 \cdot \vec{v}_2 = 0$$

$$\Rightarrow \qquad \vec{v}_1 \cdot \vec{v}_2 = 0 \qquad\qquad \because l_1 \neq l_2,\ l_1 - l_2 \neq 0$$

$\therefore$ Eigen space corresponding to $\vec{v}_1$ and $\vec{v}_2$ are orthogonal.   ∎

Theorem 13 If an $n \times n$ matrix A is orthogonally diagonalizable then A is symmetric matrix.

Proof : Since, A is orthogonally diagonalizable, by definition, there is an orthogonal matrix P(with $P^{-1} = P^t$) and a diagonal matrix D such that

$$A = PDP^t = PDP^{-1}$$

Taking transpose throughout and using reversal law of transpose, we have

$$
\begin{aligned}
A^t &= (PDP^t)^t \\
&= (P^t)^t D^t P^t \\
&= PDP^t \\
&= A
\end{aligned}
$$

Therefore, A is symmetric matrix.

Remark : The converse of the statement is also true, thus the above theorem can be restated as :

"An $n \times n$ matrix A is orthogonally diagonalizable if and only if A is a symmetric matrix."

> **Spectral Theorem (Without Proof)**

If A is $n \times n$ symmetric matrix then A has following properties :

1. A has n real Eigen values with counting multiplicities.

2. The dimension of the Eigen space for each Eigen value of A equals to the multiplicity of l which is root of characteristic equation.

3. The Eigen spaces are mutually orthogonal i.e. Eigen vectors corresponding to different Eigen values are orthogonal.

4. A is orthogonally diagonalizable.

3.5 Process of Orthogonally Diagonalizing Matrix

Given any matrix of order $n \times n$ which is symmetric i.e. $A = A^t$.

Step I : Find out Eigen values of A with counting multiplicity.

Step II : Find out corresponding Eigen vectors of A.

Let $l_1, l_2, ..., l_k$ are Eigen values of A with counting multiplicity and $\vec{v}_1, \vec{v}_2, ..., \vec{v}_k$ are Eigen vectors of A correspond to the Eigen values $l_1, l_2, ..., l_k$.

Step III : Normalize Eigen vectors let, $\vec{u}_1, \vec{u}_2, ..., \vec{u}_k$ be the normalized vectors corresponding to $\vec{v}_1, \vec{v}_2, ..., \vec{v}_k$ (all are distinct).

Step IV : Write $A = PDP^t$.

Note that, if $\{\vec{v}_1, \vec{v}_2\}$ are the Eigen vectors corresponding to Eigen value l_1 (say) then $\{\vec{u}_1, \vec{u}_2\}$ are normalize vectors corresponding to $\{\vec{v}_1, \vec{v}_2\}$ calculated as,

$$\vec{u}_1 = \frac{\vec{v}_1}{\|\vec{v}_1\|} \text{ and } \vec{u}_2 = \vec{v}_2 - \frac{\vec{v}_1 \cdot \vec{v}_2}{\vec{v}_1 \cdot \vec{v}_1} \vec{v}_1 \text{ (i.e. Proj. of } \vec{v}_2 \text{ onto } \vec{v}_1) \quad \blacksquare$$

Example 3.28 : *Orthogonally diagonalize the matrix,*

$$A = \begin{bmatrix} 1 & -2 & 2 \\ -2 & 4 & -4 \\ 2 & -4 & 4 \end{bmatrix}.$$

Solution : Step I : The characteristic equation of A is,

$$\det (A - lI_3) = 0 \text{ gives } l^2 (l - 9) = 0$$

$$\therefore \quad l_1 = 9, \ l_2 = 0 \text{ and } l_3 = 0$$

Step II : $\vec{v}_1 = \begin{bmatrix} 1 \\ -2 \\ 2 \end{bmatrix}$ is an Eigen vector corresponding to $l_1 = 9$.

$$\left\{ \vec{v}_2 = \begin{bmatrix} 2 \\ 1 \\ 0 \end{bmatrix}, \ \vec{v}_3 = \begin{bmatrix} -2 \\ 0 \\ 1 \end{bmatrix} \right\} \text{ are the Eigen vectors corresponding to } l = 0.$$

Step III : $\vec{u}_1 = \dfrac{\vec{v}_1}{\|\vec{v}_1\|} = \dfrac{(1, -2, 2)}{\sqrt{1 + 4 + 4}} = \dfrac{1}{3}(1, -2, 2)$

$$\therefore \quad \vec{u}_1 = \begin{bmatrix} \dfrac{1}{3} \\[2mm] \dfrac{-2}{3} \\[2mm] \dfrac{2}{3} \end{bmatrix}$$

$$\vec{u}_2 = \dfrac{\vec{v}_2}{\|\vec{v}_2\|} = \dfrac{(2, 1, 0)}{\sqrt{5}} = \begin{bmatrix} \dfrac{2}{\sqrt{5}} \\[2mm] \dfrac{1}{\sqrt{5}} \\[2mm] 0 \end{bmatrix} \qquad \therefore \quad \vec{u}_2 = \begin{bmatrix} \dfrac{2}{\sqrt{5}} \\[2mm] \dfrac{1}{\sqrt{5}} \\[2mm] 0 \end{bmatrix}$$

$$\vec{u}_3 = \vec{v}_3 - \dfrac{\vec{v}_2 \cdot \vec{v}_3}{\vec{v}_2 \cdot \vec{v}_2} \vec{v}_2$$

$$= (-2, 0, 1) - \dfrac{[(2, 1, 0) \cdot (-2, 0, 1)]}{(2, 1, 0) \cdot (2, 1, 0)} (2, 1, 0)$$

$$= (-2, 0, 1) - \dfrac{[-4 + 0 + 0]}{[4 + 1 + 0]} (2, 1, 0) = (-2, 0, 1) + \left(\dfrac{8}{5}, \dfrac{4}{5}, 0 \right)$$

$$\Rightarrow \quad \vec{u}_3 = \left(-2 + \dfrac{8}{5}, \dfrac{4}{5}, 1 \right) = \left(\dfrac{-2}{5}, \dfrac{4}{5}, 1 \right)$$

$$\therefore \quad \vec{u}_3 = \dfrac{\left(\dfrac{-2}{5}, \dfrac{4}{5}, 1 \right)}{\sqrt{\dfrac{4}{25} + \dfrac{16}{25} + 1}} = \dfrac{(-2, 4, 5)}{\sqrt{45}} = \dfrac{1}{3\sqrt{5}} (-2, 4, 5)$$

$$\therefore \quad \vec{u}_3 = \begin{bmatrix} \dfrac{-2}{3\sqrt{5}} \\ \dfrac{4}{3\sqrt{5}} \\ \dfrac{5}{3\sqrt{5}} \end{bmatrix}$$

Step IV : $P = \begin{bmatrix} \dfrac{1}{3} & \dfrac{2}{\sqrt{5}} & \dfrac{-2}{3\sqrt{5}} \\ \dfrac{-2}{3} & \dfrac{1}{\sqrt{5}} & \dfrac{4}{3\sqrt{5}} \\ \dfrac{2}{3} & 0 & \dfrac{5}{3\sqrt{5}} \end{bmatrix}$ and $D = \begin{bmatrix} 1 & 0 & 0 \\ 0 & 0 & 0 \\ 0 & 0 & 0 \end{bmatrix}$

such that $A = PDP^{T}$

3.5.1 Process of Spectral Decomposition

Given symmetric matrix A of order $n \times n$. To write A into its spectral decomposition form :

Step I : Orthogonally diagonal the matrix and find P, D, P^{T}.

Step II :

$$A = PDP^{T} = \underset{1 \times n}{[\vec{u}_1\ \vec{u}_2\text{- - -}\vec{u}_n]} \underset{\substack{n \times n \\ D}}{\begin{bmatrix} \lambda_1 & & 0 \\ & \lambda_2 & \\ & & \ddots \\ 0 & & \lambda_1 \end{bmatrix}} \underset{\substack{n \times 1 \\ P^{T}}}{\begin{bmatrix} \vec{u}_1^{T} \\ \vec{u}_2^{T} \\ \vec{u}_7^{T} \end{bmatrix}}$$

Step III :

$$A = \underset{\substack{1 \times n \\ PD}}{[\lambda_1\vec{u}_1\ \ \lambda_2\vec{u}_2\text{- - -}\lambda_{2n}\vec{u}_n]} \underset{\substack{n \times 1 \\ P^{T}}}{\begin{bmatrix} \vec{u}_1^{T} \\ \vec{u}_2^{T} \\ \vec{u}_7^{T} \end{bmatrix}}$$

Step IV : $A = l_1\,\vec{u}_1\,\vec{u}_1^{t} + l_2\,\vec{u}_2\,\vec{u}_2^{t} + \ldots + l_n\,\vec{u}_n\,\vec{u}_n^{t}$

The above representation is called spectral decomposition of A.

Example 3.25 : *Construct a spectral decomposition of the matrix A*

where, $A = \begin{bmatrix} 3 & 2 & 4 \\ 2 & 0 & 2 \\ 4 & 2 & 3 \end{bmatrix}$.

Solution : Given :

Step I : The characteristic equation of A is $(l+1)^2 \cdot (l-8) = 0$

$\Rightarrow$ $l = (-1), (-1)$ and 8 are Eigen values of A.

$\therefore$ $D = \begin{bmatrix} -1 & 0 & 0 \\ 0 & -1 & 0 \\ 0 & 0 & 8 \end{bmatrix}$ is the diagonal matrix.

Step II : To find Eigen vectors corresponding to Eigen values $-1, 8$.

The Eigen vectors corresponding to $l = -1$ are

$$\left\{ \vec{v}_1 = \begin{bmatrix} 1 \\ -2 \\ 0 \end{bmatrix}, \vec{v}_2 = \begin{bmatrix} 4 \\ 2 \\ -5 \end{bmatrix} \right\}$$

The Eigen vector corresponding to $l = 8$ is $\vec{v}_3 = \begin{bmatrix} 2 \\ 1 \\ 2 \end{bmatrix}$.

Step III : To normalize the Eigen vectors

$$\vec{u}_1 = \frac{\vec{v}_1}{\|\vec{v}_1\|}, \quad \vec{u}_2 = \vec{v}_2 - \frac{\vec{v}_2 \cdot \vec{v}_1}{\vec{v}_1 \cdot \vec{v}_1} \vec{v}_1 \text{ and } \vec{u}_3 = \frac{\vec{v}_3}{\|\vec{v}_3\|}$$

$$\therefore \quad \vec{u}_1 = \frac{(1, -2, 0)}{\sqrt{5}}, \quad \vec{u}_2 = (4, 2, -5) - \frac{(4, 2, -5) \cdot (1, -2, 0)}{(1, -2, 0) \cdot (1, -2, 0)} (1, -2, 0)$$

$$\Rightarrow \vec{u}_2 = (4, 2, -5) - \frac{[4 - 4 + 0]}{(1 + 4 + 0)} (1, -2, 0) = (4, 2, -5)$$

$$\therefore \quad \vec{u}_2 = \frac{(4, 2, -5)}{\sqrt{16 + 4 + 25}} = \frac{(4, 2, -5)}{\sqrt{45}} = \left(\frac{4}{3\sqrt{5}}, \frac{2}{3\sqrt{5}}, \frac{-5}{3\sqrt{5}} \right)$$

$$\vec{u}_3 = \frac{\vec{v}_3}{\|\vec{v}_3\|} = \frac{(2, 1, 2)}{\sqrt{9}} = \left(\frac{2}{3}, \frac{1}{3}, \frac{2}{3} \right)$$

$$\therefore \quad P = [\vec{u}_1 \ \vec{u}_2 \ \vec{u}_3] = \begin{bmatrix} \dfrac{1}{\sqrt{5}} & \dfrac{4}{3\sqrt{5}} & 0 \\[3mm] \dfrac{-2}{\sqrt{5}} & \dfrac{2}{3\sqrt{5}} & \dfrac{1}{3} \\[3mm] 0 & \dfrac{-5}{3\sqrt{5}} & \dfrac{2}{3} \end{bmatrix}$$

$$\text{and } P^{-1} = P^t = \begin{bmatrix} \dfrac{1}{\sqrt{5}} & \dfrac{-2}{\sqrt{5}} & \dfrac{2}{3} \\[3mm] \dfrac{4}{3\sqrt{5}} & \dfrac{2}{3\sqrt{5}} & \dfrac{1}{3} \\[3mm] \dfrac{2}{3} & \dfrac{1}{3} & \dfrac{2}{3} \end{bmatrix}$$

Step IV : $A = PDP^t$

$$= \underbrace{\begin{bmatrix} \dfrac{1}{\sqrt{5}} & \dfrac{4}{3\sqrt{5}} & \dfrac{2}{3} \\[3mm] \dfrac{-2}{\sqrt{5}} & \dfrac{2}{3\sqrt{5}} & \dfrac{1}{3} \\[3mm] 0 & \dfrac{-5}{3\sqrt{5}} & \dfrac{2}{3} \end{bmatrix}}_{P} \underbrace{\begin{bmatrix} -1 & 0 & 0 \\ 0 & -1 & 0 \\ 0 & 0 & 8 \end{bmatrix}}_{D} \underbrace{\begin{bmatrix} \dfrac{1}{\sqrt{5}} & \dfrac{-2}{\sqrt{5}} & 0 \\[3mm] \dfrac{4}{3\sqrt{5}} & \dfrac{2}{3\sqrt{5}} & \dfrac{1}{3} \\[3mm] \dfrac{2}{3} & \dfrac{1}{3} & \dfrac{2}{3} \end{bmatrix}}_{P^t}$$

$$\because \quad A = l_1 \, \vec{u}_1 \, \vec{u}_1^{\,t} + l_2 \, \vec{u}_2 \, \vec{u}_2^{\,t} + l_3 \, \vec{u}_3 \, \vec{u}_3^{\,t}$$

$$\Rightarrow \quad A = \underbrace{(-1)}_{l_1} \underbrace{\begin{bmatrix} \dfrac{1}{\sqrt{5}} \\[3mm] \dfrac{4}{3\sqrt{5}} \\[3mm] \dfrac{2}{3} \end{bmatrix}_{3 \times 1}}_{\vec{u}_1} \underbrace{\begin{bmatrix} \dfrac{1}{\sqrt{5}} & \dfrac{4}{3\sqrt{5}} & \dfrac{2}{3} \end{bmatrix}_{1 \times 3}}_{\vec{u}_1^{\,t}} + \underbrace{(-1)}_{l_2} \underbrace{\begin{bmatrix} \dfrac{4}{3\sqrt{5}} \\[3mm] \dfrac{2}{3\sqrt{5}} \\[3mm] \dfrac{-5}{3\sqrt{5}} \end{bmatrix}_{3 \times 1}}_{\vec{u}_2}$$

$$\underbrace{\begin{bmatrix} \dfrac{4}{3\sqrt{5}} & \dfrac{2}{3\sqrt{5}} & \dfrac{-5}{3\sqrt{5}} \end{bmatrix}_{1 \times 3}}_{\vec{u}_2^{\,t}} + \underbrace{8}_{l_3} \underbrace{\begin{bmatrix} 0 \\[2mm] \dfrac{1}{3} \\[2mm] \dfrac{2}{3} \end{bmatrix}_{3 \times 1}}_{\vec{u}_3} \underbrace{\begin{bmatrix} 0 & \dfrac{1}{3} & \dfrac{2}{3} \end{bmatrix}_{1 \times 3}}_{\vec{u}_3^{\,t}}$$

Think Over It

1. How much time will require to watch all the uploaded videos on you tube till today date ?
2. What is 4 colour problem in mathematics ?

Summary

1. Dot product is commutative, distributive, scalar associative.

2. Two vectors are orthogonal if there dot product is zero.

3. If $\vec{u}$ is orthogonal to every vector in subspace W of $\mathbb{R}^n$ then $\vec{u}$ is in orthogonal complement of W.

4. $W^\perp$ is always subspace of $\mathbb{R}^n$.

5. For given matrix A of order $m \times n$ then $(\text{Row } A)^t = \text{Nul } A$ and $(\text{Col } A)^t = \text{Nul } A^t$.

6. If A is $m \times n$ matrix then A has orthonormal columns if and only if $A^t A = I$.

7. If W is any subspace of $\mathbb{R}^n$. Then every $\vec{w} \in \mathbb{R}^n$ can be written uniquely in the form of $\vec{w} = \hat{w} + \vec{z}$ where $\hat{w} \in W$ and $\vec{z} \in W^\perp$.

8. An $n \times n$ matrix A is orthogonally diagonalizable iff A is symmetric.

9. Given symmetric matrix A of order $n \times n$ then A has spectral decomposition form.

Exercise

[A] Say True or False : Justify !

1. If A is real $n \times n$ symmetric matrix then A^{-1} is also symmetric.

2. If $\vec{u}$ and $\vec{v}$ are vectors in $\mathbb{R}^n$ such that $\vec{u} \cdot \vec{v} \neq 0$ then $\vec{u}$ and $\vec{v}$ may be orthogonal.

3. $\|\vec{u} + \vec{v}\|^2 + \|\vec{u} - \vec{v}\|^2 = 2\|\vec{u}\|^2 - 2\|\vec{v}\|^2$ for every vector $\vec{u}$, $\vec{v}$ in $\mathbb{R}^n$.

4. If $\vec{u}, \vec{v} \in \mathbb{R}^n$ then $\text{dist}(\vec{u}, \vec{v}) = \|\vec{u} + \vec{v}\|$.

5. The norm of vector is sometimes negative.

[B] Multiple Choice Questions : Choose the Correct Alternative

1. The two vectors $\{\vec{u} = (4, -1), \vec{v} = (1, 4)\}$ are ......

    (a) a linearly dependent of each other

    (b) forming an orthonormal basis

    (c) perpendicular to each other

    (d) none of these

2. The dot product of two orthogonal vectors is ......

    (a) 2π (b) π

    (c) 0   (d) none of these

3. If α is scalar value, $\vec{v}$ and $\vec{w}$ are any two vectors in $\mathbb{R}^n$ then $\alpha + (\vec{v} \times \vec{w}) \cdot (\vec{v} \times \vec{w})$ is ......

    (a) vector in $\mathbb{R}^n$ (b) scalar

    (c) not defined   (d) none of these

4. If the dot product of two non-zero vectors is zero then they ......

    (a) are linearly dependent

    (b) can form an orthonormal basis

    (c) have exactly opposite directions to each other

    (d) none of these

5. Orthogonal Projection of a $\vec{u}$ onto $\vec{v}$ is ......

    (a) $\dfrac{(\vec{u} \cdot \vec{v})\,\vec{u}}{\vec{u} \cdot \vec{u}}$ (b) $\dfrac{(\vec{u} \cdot \vec{v})\,\vec{v}}{\vec{v} \cdot \vec{v}}$

    (c) $\dfrac{\vec{u}}{\vec{v} \cdot \vec{v}}$ (d) none of these

6. If $\vec{u} \cdot \vec{u} = \vec{v} \cdot \vec{v}$ then $(\vec{u} + \vec{v}) \cdot (\vec{u} - \vec{v}) = $ ......

    (a) positive   (b) negative

    (c) 0   (d) none of these

7. The vectors $(2, 3, -4)$ and (a, b, c) are perpendicular if ......

   (a) $a = 2, b = 3, c = -4$ (b) $a = 4, b = 4, c = 5$

   (c) $a = 4, b = 4, c = -5$ (d) none of these

8. If A is $n \times n$ matrix then A is symmetric matrix if, ...

   (a) $A = A^t$ (b) $A = A^{-1}$

   (c) $A = A^2$ (d) None of these

9. If A is symmetric real matrix then A has ......

   (a) red eigen values   (b) all eigen values are complex

   (c) all eigen values are integer   (d) none of these

10. If $\vec{u}$ and $\vec{v}$ are vectors in $\mathbb{R}^n$ then $\|\vec{u} - \vec{v}\|$ gives ......

   (a) distance between $\vec{u}$ and $\vec{v}$ (b) angle between $\vec{u}$ and $\vec{v}$

   (c) always vector in $\mathbb{R}^n$ (d) none of these

[C] Theory Questions :

1. If $\vec{u}$ and $\vec{v}$ are any vectors in $\mathbb{R}^n$ then show that

   (a) $\|\vec{u} + \vec{v}\|^2 + \|\vec{u} - \vec{v}\|^2 = 2\,(\|\vec{u}\|^2 + \|\vec{v}\|^2)$

   (b) $\|\vec{u} + \vec{v}\|^2 - \|\vec{u} - \vec{v}\|^2 = 4\,\vec{u} \cdot \vec{v}$

2. If $\vec{u}$ and $\vec{v}$ are any vectors in $\mathbb{R}^n$ then show that $|\vec{u} \cdot \vec{v}| \leq \|\vec{u}\|\,\|\vec{v}\|$.
   (Cauchy - Schwartz inequality)

3. State and prove Pythagorean for the vectors in $\mathbb{R}^n$.

4. State and prove law of parallelogram of addition vectors in $\mathbb{R}^n$.

5. If possible, diagonalize the matrix $A = \begin{bmatrix} 6 & -2 & -1 \\ -2 & 6 & -1 \\ -1 & -1 & 5 \end{bmatrix}$.

6. Show every invertible symmetric matrix A of order $n \times n$ is also symmetric.

7. Show that every real symmetric matrix A of order $n \times n$ has real eigen values.

8. Show that every real symmetric matrix A is orthogonally diagonalizable.

9. Show that given two vectors $\vec{u}$ and $\vec{v}$ in $\mathbb{R}^n$ then $-1 \le \dfrac{\vec{u} \cdot \vec{v}}{\|\vec{u}\| \, \|\vec{v}\|} \le +1$.

   (**Hint :** Use the property of cosine angle between two vectors)

10. If $\vec{u}, \vec{v} \in \mathbb{R}^n$ are orthogonal vectors then $\|\vec{u} + \vec{v}\|^2 + \|\vec{u}\|^2 + \|\vec{v}\|^2$.

11. Show that the zero vector is the only vector in $\mathbb{R}^n$ which is orthogonal to every vector in $\mathbb{R}^n$.

12. For all real values of a, b and θ, show that

$$[a \cos \theta + b \sin \theta]^2 \le a^2 + b^2$$

13. Show for any two vectors $\vec{u}$ and $\vec{v}$ in $\mathbb{R}^n$

   (a) $\vec{u} \cdot \vec{v} = 0$ if and only if $\|\vec{u} + \vec{v}\| = \|\vec{u} - \vec{v}\|$

   (b) $(\vec{u} + \vec{v}) \cdot (\vec{u} - \vec{v}) = 0$ if and only if $\|\vec{u}\| = \|\vec{v}\|$.

14. Let $S = \{\vec{v}_1, \vec{v}_2, \ldots, \vec{v}_n\}$ be set of non-zero vectors in $\mathbb{R}^n$ if all pairs of distinct vectors of S are orthogonal then show that S is linearly independent.

15. Show that, if A is $n \times n$ matrix having orthogonal columns then determine of A is ± 1.

[D] Numerical Problems :

1. Suppose $\vec{u} \cdot \vec{v} = 1$, $\vec{v} \cdot \vec{w} = 3$, $\vec{u} \cdot \vec{w} = 2$, $\|\vec{u}\| = 3$, $\|\vec{v}\| = 2$ and $\|\vec{w}\| = 1$ then evaluate (a) $(\vec{u} - \vec{v} - 2\vec{w}) \cdot (4\vec{u} + \vec{v})$, (b) $\|2\vec{w} - \vec{u}\|$.

2. Find the norm of each vector and distance between $\vec{u} = (1, 1, -1)$ and $\vec{w} = (-1, 1, 0)$.

3. Find two vectors in $\mathbb{R}^4$ of norm 1 that are orthogonal to following 3 vectors $\vec{u} = (2, 1, -4, 0)$, $\vec{v} = (-1, -1, 2, 2)$ and $\vec{w} = (3, 2, 5, 4)$.

4. Orthogonally diagonalize the matrix $A = \begin{bmatrix} 3 & -2 & 4 \\ -2 & 6 & 2 \\ 4 & 2 & 3 \end{bmatrix}$.

5. Given two vectors $\vec{u}$ and $\vec{v}$ in $\mathbb{R}^n$ then the angle between then is given by $\theta = \cos^{-1}\left[\dfrac{\vec{u} \cdot \vec{v}}{\|\vec{u}\| \, \|\vec{v}\|}\right]$. Find the angle between two vectors $\vec{u} = (1, 1, 0, 1)$ and $\vec{v} = (1, 0, 2, -1)$ in $\mathbb{R}^4$.

6. Show that $S = \{\vec{u}_1, \vec{u}_2, \vec{u}_3\}$ is an orthogonal set where,

$$\vec{u}_1 = \begin{bmatrix} 3 \\ 1 \\ 1 \end{bmatrix}, \quad \vec{u}_2 = \begin{bmatrix} -1 \\ 2 \\ 1 \end{bmatrix} \text{ and } \vec{u}_3 = \begin{bmatrix} \frac{-1}{2} \\ -2 \\ \frac{7}{2} \end{bmatrix}.$$

7. If $S = \{\vec{u}_1, \vec{u}_2, \vec{u}_3\}$ where $\vec{u}_1 = \begin{bmatrix} 3 \\ 1 \\ 1 \end{bmatrix}$, $\vec{u}_2 = \begin{bmatrix} -1 \\ 2 \\ 1 \end{bmatrix}$, $\vec{u}_3 = \begin{bmatrix} \frac{-1}{2} \\ -2 \\ \frac{7}{2} \end{bmatrix}$,

$\vec{w} = \begin{bmatrix} 6 \\ 1 \\ -8 \end{bmatrix}$. Express w as linearly combination of vectors in S.

8. Let $\vec{w} = \begin{bmatrix} 7 \\ 6 \end{bmatrix}$ and $\vec{u} = \begin{bmatrix} 4 \\ 2 \end{bmatrix}$. Find the orthogonal projection of $\vec{w}$ onto $\vec{u}$. Then write $\vec{w}$ as the sum of two orthogonal vectors, one in span $\{\vec{u}\}$ and one orthogonal to $\vec{u}$.

9. Let $U = \begin{bmatrix} \frac{1}{\sqrt{2}} & \frac{2}{3} \\ \frac{1}{\sqrt{2}} & \frac{-2}{3} \\ 0 & \frac{1}{3} \end{bmatrix}$ and $X = \begin{bmatrix} \sqrt{2} \\ 3 \end{bmatrix}$ verify that $\|UX\| = \|x\|$.

10. If $\vec{u}$ and $\vec{v}$ are orthogonal vectors in $\mathbb{R}^n$ such that, $\|\vec{u}\| = 1$, $\|\vec{v}\| = 1$, then show that :

(a) $\|\vec{u} + \vec{v}\|^2 + \|\vec{u} - \vec{v}\|^2 = 4$

(b) $\|\vec{u} + \vec{v}\| = \sqrt{2}$

11. Show that, if A is $n \times n$ matrix then adj. A is also symmetric matrix.

    (**Hint :** Use the formula adj $(A) = A^{-1}$ det. A)

12. Show that, if A is orthogonally diagonalizable then A^2 is also orthogonally diagonalizable.

13. Let $X = \begin{bmatrix} x_1 \\ x_2 \end{bmatrix}$. Compute $X^t AX$ in the following matrices :

    (a) $A = \begin{bmatrix} 4 & 0 \\ 0 & 3 \end{bmatrix}$
    (b) $A = \begin{bmatrix} 3 & -2 \\ -2 & 7 \end{bmatrix}$

14. Find the values of k for which $\{\vec{u}_1 = (2, 1, 3), \vec{u}_2 = (1, 7, k)\}$ are orthogonal.

15. Show that for any real number α, vectors $\vec{u}$ and $\vec{v}$ in $\mathbb{R}^n$; $\vec{u} \cdot \vec{v} = 0$ if and only if $\|\vec{u} + \alpha\vec{v}\| \geq \|\vec{u}\|$, for all real α.

Answers

[A] (1) True   (2) False   (3) False   (4) False   (5) False

[B] (1) - (c)   (2) - (c)   (3) - (b)   (4) - (b)   (5) - (b)   (6) - (c)   (7) - (b)
    (8) - (a)   (9) - (a)   (10) - (a)

[D] 1.   (a) 13   (b) $\sqrt{5}$

2. $\|\vec{u}\| = \sqrt{3}$, $\|\vec{v}\| = \sqrt{5}$, $d(\vec{u}, \vec{v}) = \|\vec{u} - \vec{v}\| = \sqrt{13}$

3. $\pm \dfrac{1}{57} (-34, 44, -6, 11)$

4. $A = PDP^{-1}$ where, $D = \begin{bmatrix} 7 & 0 & 0 \\ 0 & 7 & 0 \\ 0 & 0 & -2 \end{bmatrix}$ and $P = \begin{bmatrix} \dfrac{1}{\sqrt{2}} & \dfrac{-1}{\sqrt{18}} & \dfrac{-2}{3} \\ 0 & \dfrac{4}{\sqrt{18}} & \dfrac{-1}{3} \\ \dfrac{1}{\sqrt{2}} & \dfrac{1}{\sqrt{18}} & \dfrac{2}{3} \end{bmatrix}$

5. $\theta = \dfrac{\pi}{2}$

7. $\vec{w} = 1\vec{u}_1 - 2\vec{u}_2 - 2\vec{u}_3$

8. Orthogonal $\text{Proj}_u \vec{w} = \begin{bmatrix} 8 \\ 4 \end{bmatrix}$.

$\vec{w} = \begin{bmatrix} 8 \\ 4 \end{bmatrix} + \begin{bmatrix} -1 \\ 2 \end{bmatrix}$ where $\begin{bmatrix} -1 \\ 2 \end{bmatrix} = \vec{w} - \text{Proj}_u \vec{w}$

13. (a) $X^T AX = 4x_1^2 + 3x_2^2$

(b) $X^T AX = 3x_1^2 - 4x_1x_2 + 7x_2^2$

14. $k = -3$

Chapter 4...
Quadratic Forms and Geometry of Vectorspace

David Hilbert

David Hilbert was a German mathematician and one of the most influential and universal mathematicians of the 19th and early 20th centuries. Hilbert discovered and developed a broad range of fundamental ideas in many areas, including invariant theory, the calculus of variations, commutative algebra, algebraic number theory, the foundations of geometry.

4.1 Introduction

Till now, we have focused mainly on machine, vectors their properties and Geometrical aspect for various definitions like determinant, norm, orthogonal vectors, orthonormal vectors, projection of one vector onto other vector etc. and still we have studied linear equation form and there properties.

Now we going to focus little bit on quadratic forms specially of symmetric matrices which are useful in engineering (image, processing, design criteria and optimization). Along with these algebraic calculations now the real life objects like line segment, cube, pyramidal structure and other solid objects can be visualize by sets of vectors having interesting applications in Computer graphics.

4.2 Quadratic Forms

Definition : A quadratic form on $\mathbb{R}^n$ is a function Q defined on $\mathbb{R}^n$ whose value at a vector $X \in \mathbb{R}^n$ is given by, $Q(X) = X^t AX$, when A is $n \times n$ symmetric matrix.

Note :

(1)   A is called matrix of the Quadratic form.

(2)   Quadratic form each team must be of degree 2.

(3)   As A as symmetric matrix $X^t AX = X^t A^t X = (X^t AX)^t$, $X \in \mathbb{R}^n$.

4.1.1 Quadratic forms for 2 × 2 Symmetric Matrices

If $Q : \mathbb{R}^2 \longrightarrow \mathbb{R}$ is a quadratic function then for any 2 × 2 **symmetric matrix** A and $X = \begin{bmatrix} x_1 \\ x_2 \end{bmatrix} \in \mathbb{R}^2$. The quadratic equation for the matrix A is given by;

$$Q(X) = X^t AX = [x_1 \ x_2] \underset{A}{\begin{bmatrix} a_{11} & a_{12} \\ a_{21} & a_{22} \end{bmatrix}} \underset{X}{\begin{bmatrix} x_1 \\ x_2 \end{bmatrix}}$$

$$\underset{X^t}{}$$

$$= a_{11} x_1^2 + a_{12} x_1 x_2 + a_{21} x_2 x_1 + a_{22} x_2^2$$

$\therefore$ A is symmetric matrix $\Rightarrow a_{12} = a_{21}$

$\Rightarrow \ Q(X) = a_{11} x_1^2 + a_{12} x_1 x_2 + a_{12} x_2 x_1 + a_{22} x_1^2$

$\therefore \ Q(X) = a_{11} x_1^2 + 2a_{12} x_1 x_2 + a_{22} x_1^2$

The above equation tells us that, the co-efficients of x_1^2 and x_2^2 go on the diagonal of A and the coefficient of $x_i x_j$ for $i \neq j$ is nothing but $(a_{ij} + a_{ji}) = 2 \ a_{ij}$ as in symmetric matrix $a_{ij} = a_{ji}$ hence for symmetric matrix the coefficient of $x_i x_j$, $i \neq j$ is always even.

Example 4.1 : *Let* $X = \begin{bmatrix} x_1 \\ x_2 \end{bmatrix}$ *compute* $Q(X)$ *for the following symmetric matrices.*

$$A = \begin{bmatrix} 5 & -5 \\ -5 & 1 \end{bmatrix}, \quad B = \begin{bmatrix} 5 & 0 \\ 0 & 10 \end{bmatrix}, \quad C = \begin{bmatrix} 3 & -2 \\ -2 & 7 \end{bmatrix}$$

Given : $Q : \mathbb{R}^2 \longrightarrow \mathbb{R}$ by $Q(X) = X^t AX$ Where, $X = \begin{bmatrix} x_1 \\ x_2 \end{bmatrix} \in \mathbb{R}^2$

(i) $A = \begin{bmatrix} 5 & -5 \\ -5 & 1 \end{bmatrix} \Rightarrow Q(X) = [x_1 \ x_2] \begin{bmatrix} 5 & -5 \\ -5 & 1 \end{bmatrix} \begin{bmatrix} x_1 \\ x_2 \end{bmatrix}$

$\Rightarrow \ Q(X) = 5x_1^2 - 10 \ x_1 x_2 + x_2^2$

(ii) $B = \begin{bmatrix} 5 & 0 \\ 0 & 10 \end{bmatrix} \Rightarrow Q(X) = [x_1 \ x_2] \begin{bmatrix} 5 & 0 \\ 0 & 10 \end{bmatrix} \begin{bmatrix} x_1 \\ x_2 \end{bmatrix}$

$\Rightarrow \ Q(X) = 5x_1^2 - 0 \ x_1 x_2 + 10x_2^2 \qquad \therefore \ Q(X) = 5x_1^2 + 10x_2^2$

(iii) $C = \begin{bmatrix} 3 & -2 \\ -2 & 7 \end{bmatrix} \Rightarrow Q(X) = [x_1 \ x_2] \begin{bmatrix} 3 & -2 \\ -2 & 7 \end{bmatrix} \begin{bmatrix} x_1 \\ x_2 \end{bmatrix}$

$\Rightarrow \ Q(X) = 3x_1^2 - 4 \ x_1 x_2 + 7x_2^2$

Example 4.2 : *If* $A \begin{bmatrix} 1 & 2 \\ 3 & 4 \end{bmatrix}$ *Compute* $Q(X)$ *for* $x \in \mathbb{R}^2$.

Given : $A = \begin{bmatrix} 1 & 2 \\ 3 & 4 \end{bmatrix}$ which is not symmetric matrix

$\therefore \quad Q(X) = [x_1 \ x_2] \begin{bmatrix} a_{11} & a_{12} \\ a_{21} & a_{22} \end{bmatrix} \begin{bmatrix} x_1 \\ x_2 \end{bmatrix} = a_{11} x_1^2 + a_{12} x_1 x_2 + a_{21} x_2 x_1 + a_{22} x_2^2$

$\Rightarrow Q(X) = [x_1 \ x_2] \begin{bmatrix} 1 & 2 \\ 3 & 4 \end{bmatrix} \begin{bmatrix} x_1 \\ x_2 \end{bmatrix} = 1 x_1^2 + 2 x_1 x_2 + 3 x_2 x_1 + 4 x_2^2$

$\therefore \quad Q(X) = x_1^2 + 5 x_1 x_2 + 4 x_2^2$

Theorem 1 : If A is n $\times$ n real symmetric matrix and $Q(X) = X^t A X$ is Quadratic form then it is unique.

Proof : let, $Q(X) = X^t A X$ be the quadratic form for the real symmetric n $\times$ n matrix.

To show that, the above form is unique.

If $\qquad Q(X) = X^t B X,$ $\qquad\qquad\qquad\qquad \forall X \in \mathbb{R}^2$

$\Rightarrow$ To show that, $X^t A X = X^t B X,$ $\qquad\qquad \forall X \in \mathbb{R}^2$

$\qquad$ Where, $A = A^t$ and $B = B^t$

$\because X^t A X = \sum_{i,\,j=1}^{n} A_{ij} x_i x_j = \sum_{i,\,j=1}^{n} B_{ij} x_i x_j$

$\Rightarrow \sum_{i,\,j=1}^{n} A_{ij} x_i x_j = \sum_{i,\,j=1}^{n} B_{ij} x_i x_j$

We know that, two polynomial are same $\Leftrightarrow$ there coefficients of same degree variables are equal.

$\therefore \quad A_{ij} = B_{ij} \ ; \ \forall \ i, j = 1, 2, \ ..... \ n$

hence, $\quad Q(X) = X^t A X$ is a unique representation. ∎

4.1.2 Quadratic forms for 3 $\times$ 3 Symmetric Matrices

If $Q : \mathbb{R}^3 \longrightarrow \mathbb{R}$ is a quadratic function then for any 3 $\times$ 3 symmetric matrix A and X = **Error!**. Then quadratic equation for the matrix A is given by,

$$Q(X) = \underset{X^t}{\underbrace{X^t A X = [x_1 \ x_2 \ x_3]}} \underset{A}{\underbrace{\begin{bmatrix} a_{11} & a_{12} & a_{13} \\ a_{21} & a_{22} & a_{23} \\ a_{31} & a_{32} & a_{33} \end{bmatrix}}} \underset{X}{\underbrace{\begin{bmatrix} x_1 \\ x_2 \\ x_3 \end{bmatrix}}}$$

$$= a_{11} x_1^2 + a_{12} x_1 x_2 + a_{13} x_1 x_3 + a_{21} x_2 x_1 + a_{22} x_2^2 + a_{23} x_2 x_3 +$$
$$a_{31} x_3 x_1 + a_{32} x_3 x_2 + a_{33} x_3^2 .$$

$\because$ A is Symmetric matrix $\Rightarrow a_{ij} = a_{ji}$; $i \neq j$

$$\Rightarrow \ Q(x) = a_{11} x_1^2 + 2a_{12} x_1 x_2 + 2a_{13} x_1 x_3 + 2a_{23} x_2 x_3 + a_{22} x_2^2 + a_{33} x_3^2$$

Example 4.3 : *Find out the quadratic forms for the following matrices.*

$$A = \begin{bmatrix} 1 & 4 & 5 \\ 4 & 2 & 6 \\ 5 & 6 & 3 \end{bmatrix}, \quad B = \begin{bmatrix} 3 & -1 & 0 \\ -1 & 2 & -1 \\ 0 & 1 & 3 \end{bmatrix}$$

Solution : (i) $A = \begin{bmatrix} 1 & 4 & 5 \\ 4 & 2 & 6 \\ 5 & 6 & 3 \end{bmatrix}$ and $X = \begin{bmatrix} x_1 \\ x_2 \\ x_3 \end{bmatrix} \in \mathbb{R}^3$

$\therefore \quad Q(X) = [x_1 \ x_2 \ x_3] \ A \begin{bmatrix} x_1 \\ x_2 \\ x_3 \end{bmatrix}$

$\Rightarrow \quad Q(X) = [x_1 \ x_2 \ x_3] \begin{bmatrix} 1 & 4 & 5 \\ 4 & 2 & 6 \\ 5 & 6 & 3 \end{bmatrix} \begin{bmatrix} x_1 \\ x_2 \\ x_3 \end{bmatrix}$

$\Rightarrow \quad Q(X) = 1\,x_1^2 + 2(4)\,x_1 x_2 + 2(5)\,x_1 x_3 + 2(6)\,x_2 x_3 + 2\,x_2^2 + 3\,x_3^2$

$\Rightarrow \quad Q(X) = x_1^2 + 2x_2^2 + 3x_3^2 + 8x_1x_2 + 10x_1x_3 + 12x_2x_3$

(ii) $A \begin{bmatrix} 3 & -1 & 0 \\ -1 & 2 & -1 \\ 0 & 1 & 3 \end{bmatrix}$ and $x = \begin{bmatrix} x_1 \\ x_2 \\ x_3 \end{bmatrix} \in \mathbb{R}^3$

$\therefore \quad Q(X) = 3\,x_1^2 + 2\,x_2^2 + 3\,x_3^2 - 2\,x_1 x_2 + 0x_1 x_3 - 2x_2 x_3$

$\therefore \quad Q(X) = 3\,x_1^2 + 2\,x_2^2 + 3\,x_3^2 - 2\,x_1 x_2 - 2x_2 x_3.$

4.1.3 Change of Variables

We know that, if A is Symmetric matrix then their exist orthogonal matrix P and diagonal matrix D containing eigen values of A such that $A = P^{-1} DP$ is diagonalizable.

Now given vector $X \in \mathbb{R}^n$, we consider new vector $y = P^{-1} X$ called as co-ordinate vector of X relative to the basis of $\mathbb{R}^n$ determined by the columns of P.

AS, $Q(X)$ for the symmetric matrix A at $X \in \mathbb{R}^n$ is given by,

$Q(X) = X^t\, AX$

$\qquad = (PY)^t\, A\,(PY) \qquad - \; Y = P^{-1}X \qquad\qquad\qquad \therefore X = PY$

$\qquad = Y^t\,(P^t\, AP)\, Y$

$\qquad = Y^t\, DY \qquad\qquad - \; P^t\, AP = D$

Where, D is diagonal matrix having eigenvalues of A on the diagonal.

Example 4.4 : *Find the Quadratic form of the matrix A by change of variable that transforms the quadratic form into quadratic form with no cross-product them. Where,*

$$A = \begin{bmatrix} 3 & 2 & 4 \\ 2 & 0 & 2 \\ 4 & 2 & 3 \end{bmatrix}$$

Solution :

Step (I) : Given $\qquad A = \begin{bmatrix} 3 & 2 & 4 \\ 2 & 0 & 2 \\ 4 & 2 & 3 \end{bmatrix}$

$\therefore \quad Q(X) = 3\, x_1^2 + 0\, x_2^2 + 3\, x_3^2 + 4\, x_1\, x_2 + 8\, x_1\, x_3 + 4\, x_2\, x_3$

Step (II) : The eigen values of A are $\lambda = -1, -1, -8$

$\therefore \quad D = \begin{bmatrix} -1 & 0 & 0 \\ 0 & -1 & 0 \\ 0 & 0 & 8 \end{bmatrix} \quad$ With $\; Y = P^{-1}X = \begin{bmatrix} y_1 \\ y_2 \\ y_3 \end{bmatrix} \in \mathbb{R}^3$

Step (III) : $\qquad Q(X) \;=\; X^t\, AX = (PY)^t\, A\,(PY) \quad - X = PY$

$\qquad\qquad\qquad\qquad\quad = Y^t\,(P^t\, AP)\, Y$

$\qquad\qquad\qquad\qquad\quad = Y^t\, DY \qquad\qquad\qquad - P^T\, AP = D$

$\therefore \quad Q(X) = [y_1 \; y_2 \; y_3] \begin{bmatrix} -1 & 0 & 0 \\ 0 & -1 & 0 \\ 0 & 0 & 8 \end{bmatrix} \begin{bmatrix} y_1 \\ y_2 \\ y_3 \end{bmatrix}$

$\Rightarrow \quad Q(X) = -\,y_1^2 - y_2^2 + 8y_3^2 = 3x_1^2 + 3x_3^2 + 4x_1\, x_2 + 8x_1x_3 + 4x_2x_3$

Where $Y = P^{-1}\, X$ and it does not contain any cross-product $x_i\, x_j\ (i \neq j)$ terms.

Example 4.5 : *By using change of variable $X = PY$. Find new Quadratic form of the following matrix A which doesn't contain any Cross product term, also find Y if;*

$$X = \begin{bmatrix} 1 \\ 2 \end{bmatrix} \text{ and } A = \begin{bmatrix} 1 & -4 \\ -4 & -5 \end{bmatrix}$$

Solution : Step (I) : Given $A = \begin{bmatrix} 1 & -4 \\ -4 & -5 \end{bmatrix}$ and $X = \begin{bmatrix} x_1 \\ x_2 \end{bmatrix} \in \mathbb{R}^2$

$Q(X) = x_1^2 - 8x_1x_2 - 5x_2^2$ which contain cross product term x_1x_2.

Step (II) : The eigen values of matrix A are $\lambda = 3, -7$

$\therefore \quad D = \begin{bmatrix} 3 & 0 \\ 0 & -7 \end{bmatrix} \qquad$ with $X = PY \in \mathbb{R}^2$

Step (III) : $\qquad Q(X) = X^t AX$

$$= (PY)^t A\,(PY)$$

$$= Y^t\,(P^tAP)Y \qquad\qquad -(AB)^t = B^t A^t$$

$$= Y^t DY \qquad\qquad -P^t AP = D$$

$\therefore \quad Q(X) = [y_1 \; y_2] \begin{bmatrix} 3 & 0 \\ 0 & -7 \end{bmatrix} \begin{bmatrix} y_1 \\ y_2 \end{bmatrix} = 3y_1^2 - 7y_2^2$

$\Rightarrow \quad x_1^2 - 8x_1 x_2 - 5x_2^2 = 3y_1^2 - 7y_2^2$

Step (IV) $X = \begin{bmatrix} 1 \\ 2 \end{bmatrix} \qquad \therefore \; 3y_1^2 - 7 y_2^2 = (1)^2 - 8(1)(2) - 5(2)^2$

$$= 1 - 16 - 20 = -35$$

$\Rightarrow \quad 3y_1^2 - 7 y_1^2 = -35.$

The above procedure of finding Quadratic equation of symmetric matrix by change of variable is nothing but, "Principal Axes Theorem".

| **Theorem 2** | **Principal Axes Theorem**

If A is $n \times n$ symmetric matrix then there is an orthogonal change of variable $X = PY$, that transforms the quadratic equation $X^t AX$ into a quadratic form $Y^t DY$ with no cross product terms.

Proof : Let, A is any $n \times n$ symmetric matrix then its quadratic equation form is given by.

$$Q(X) = X^t AX \qquad\qquad \dots (1)$$

now, $X = PY$ where, P is an orthogonal matrix

$\therefore \quad$ From equation (1)

$$Q(X) = (PY)^t A(PY)$$

$$= Y^t\,(P^t AP)\,Y \qquad\qquad \dots (AB)^t = B^t A^t$$

$$= Y^t DY \qquad\qquad \dots P^t AP = D$$

$\therefore \quad Q(X) = Q(Y) = Y^t DY$

Where, D is diagonal matrix i.e. matrix with non-diagonal entries are zero $\Rightarrow$ Q (Y) does not contain any Cross product term.

and $\qquad$ $Q(Y) = a_{11}\, y_1^2 + a_{22}\, y_2^2 + \ldots\ldots + a_{nn}\, y_n^2.$ ∎

Note that, the quadratic forms in general encode the so called "quadratic surfaces" such as ellipse, hyperbolic paraboloids etc. Due to Principal Axes Theorem there are no more Cross-product terms in quadratic forms of symmetric matrices.

Example 4.6 : *Find the change of variable that removes the cross-product term from the equation* $5x_1^2 - 10\,x_1x_2 + x_2^2 = 10.$

Solution : Step (I) : Given $Q(X) = 5x_1^2 - 10x_1x_2 + x_2^2 = 10$ then the Corresponding symmetric matrix A is given by,

$$A = \begin{bmatrix} 5 & -5 \\ -5 & 1 \end{bmatrix}$$

Whose eigen values are $\lambda = 3 + \sqrt{29}$, $3 - \sqrt{29}$ and Corresponding eigenvectors are given by,

$$\vec{u_1} = \begin{bmatrix} 1/5 + 1/5\,(-\,3 -\sqrt{29}\,) \\ 1 \end{bmatrix} \qquad \text{For } \lambda_1 = 3 + \sqrt{29}$$

$$\text{and } \vec{u_2} = \begin{bmatrix} 1/5 + 1/5\,(\sqrt{29} - 3) \\ 1 \end{bmatrix} \qquad \text{For } \lambda_2 = 3 - \sqrt{29}$$

Step (II) : Define $P = [\vec{u}_1\ \vec{u}_2]$

Then P is orthogonally diagonalizes the matrix A, so that by change of variable $X = PY$ produces the Quadratic form

$$Y^t\,DY = [y_1\ y_2] \begin{bmatrix} 3 + \sqrt{29} & 0 \\ 0 & 3 - \sqrt{29} \end{bmatrix} \begin{bmatrix} y_1 \\ y_2 \end{bmatrix}$$

$$\Rightarrow \quad Q(X) = Q(y) = (3 + \sqrt{29}\,)\, y_1^2 + (3 - \sqrt{29}\,)\, y_2^2.$$

4.1.4 Classification of Quadratic Forms

Given a symmetric matrix A of order $n \times n$ its quadratic form were represented by $Q(X) = X^t\,AX$ having real value and domain $\mathbb{R}^n$.

We have seen in the example 4.5, how the value of $Q(X)$ can be obtained at given $X \in \mathbb{R}^2$. This value may be positive negative or may be zero depends upon which the Quadratic form $Q(X)$ has been classified.

Definition : If $Q(X)$ is quadratic form of a symmetric matrix A of order n×n then

(i)   It is positive definite if $Q(X) > 0$, $\forall X \in \mathbb{R}^n\ \{\vec{0}\}$ i.e. $X \neq \vec{0}$

(ii) It is negative definite if $Q(X) < 0$, $\forall X \neq \vec{0}$ in $\mathbb{R}^n$

(iii) It is indefinite if $Q(X)$ assumes both positive and negative values for $X \in \mathbb{R}^n$.

Note that, $Q(X)$ is also called positive semidefinite if $Q(X) \geq 0$ $\forall X \in \mathbb{R}^n$ while negative semidefinite if $Q(X) \leq 0$ $\forall X \in \mathbb{R}^n$. The classification of some quadratic forms of symmetric matrices of order $n \times n$ can be done by using the properties of an eigenvalues as follows :

$\boxed{\textbf{Theorem 3}}$ Let A be an $n \times n$ symmetric matrix having Quadratic form $Q(X) = X^t AX$ then :

(a) $Q(X)$ is positive definite if and only if the eigenvalues of A are all positive.

(b) $Q(X)$ is negative definite if and only if the eigencvalues of A are all negative.

(c) $Q(X)$ is indefinite if and only if A has both positive and negative eigenvalues.

Proof : Given a Symmetric matrix A of order $n \times n$ having quadratic from;

$$Q(X) = X^t AX \qquad \qquad \text{... (I)}$$

Now by using, "Principal Axes Therom" there exist an orthogonal change of variable;

$$X = PY \qquad \qquad \text{... (II)}$$

Using equation (I) and (II)

We get,

$$Q(X) = X^t AX = Y^t DY$$

Where, D is diagonal matrix with entries on the diagonal are eigenvalues of matrix A.

Let $\lambda_1, \lambda_2, \ldots \lambda_n$ be the eigen values of A

$$\Rightarrow Q(X) = Y^t DY = \lambda_1 y_1^2 + \lambda_2 y_2^2 + \ldots + \lambda_n y_n^2 \qquad \text{... (III)}$$

From equation (II) $\Rightarrow Y = P^{-1}X$ $\qquad$ ... (IV)

$\because$ P is invertible there is one-to-one correspondence between all non-zero X and non-zero Y.

Hence, $Q(X)$ coincides with $\lambda_1 y_1^2 + \lambda_2 y_1^2 + \ldots + \lambda_n y_2^2$. and as $y_1^2, y_2^2, \ldots, y_n^2$ all are non negative the signs of $Q(X)$ are controlled by the signs of $\lambda_1, \lambda_2, \ldots, \lambda_n$.

Case (I) : All $\lambda_i > 0 \Leftrightarrow Q(X) > 0 \ \forall X \neq 0$

i.e. $Q(X)$ is positive definite $\Leftrightarrow \lambda_i > 0$; $i = 1, 2, \ldots, n$

Case (II) : All $\lambda_i < 0 \Leftrightarrow Q(X) < 0 \ \forall X \neq 0$

i.e. $Q(X)$ is negative definite $\Leftrightarrow \lambda i < 0$ for $i = 1, 2, \ldots, n$

Otherwise $Q(X) > 0$ or $Q(X) < 0$ for $X \in \mathbb{R}^n$, i.e. $Q(X)$ is indefinite.

Thus to check whether to check $Q(X)$ is positive definite or negative definite or indefinite, there is no need to check signs of $Q(X)$ for each X in $\mathbb{R}^n$. One can just check the signs of eigevalues of symmetric matrix A of order n × n.

Example 4.7 : *Check whether,* $Q(X) = 3x_1^2 - 4\,x_1x_2 + 5x_2^2$ *is positive definite or negative definite or indefinite.*

Solution : Given $Q(X) = 3x_1^2 - 4\,x_1x_2 + 5x_2^2,\ x = \begin{bmatrix} x_1 \\ x_2 \end{bmatrix} \in \mathbb{R}^2$

Step (I) : The Symmetric matrix A of order 2×2 is given by,

$A = \begin{bmatrix} 3 & -2 \\ -2 & 5 \end{bmatrix}$ whose eigen values are $\lambda_1 = 4 - \sqrt{5}$ and

$\lambda_2 = 4 + \sqrt{5}$

Step (II) : $\lambda_i = 4 - \sqrt{5} > 0$ also $\lambda_2 = 4 + \sqrt{5} > 0$

$\therefore$ By the property $Q(X)$ is positive definite.

Example 4.8 : *Check whether* $Q(X) = 3x_1^2 = 2x_2^2 + x_3^2 + 4x_1x_2 + 4x_2x_3$ *is positive definite or negative definite or indefinite.*

Solution : Given $Q(X) = 3x_1^2 + 2x_2^2 + x_3^2 + 4x_1x_2 + 4x_2x_3,$

Where, $X = \begin{bmatrix} x_1 \\ x_2 \\ x_3 \end{bmatrix} \in \mathbb{R}^3$

Step (I) : The symmetric matrix A of order 3×3 is given by comparing with

$$Q(X) = a_{11}\,x_1^2 + a_{22}\,x_2^2 + a_{33}\,x_3^2 + (a_{12} + a_{21})\,x_1x_2 + (a_{23} + a_{32})$$
$$x_2x_3 + (a_{13} + a_{31})\,x_1x_3$$

$\Rightarrow a_{11} = 3,\ a_{22} = 2,\ a_{33} = 1,\ a_{12} = a_{21} = 2,\ a_{23} = a_{32} = 2,\ a_{13} = a_{31} = 0$

$\therefore$ $A = \begin{bmatrix} 3 & 2 & 0 \\ 2 & 2 & 2 \\ 0 & 2 & 1 \end{bmatrix}$

Step (II) Now the eigen values of $A = \begin{bmatrix} 3 & 2 & 0 \\ 2 & 2 & 2 \\ 0 & 2 & 1 \end{bmatrix}$ are $\lambda_1 = 5$, $\lambda_2 = 2$

and $\lambda_3 = -1$ here not all eigen values are positive or all are negative

$\Rightarrow$ $Q(X)$ is on indefinite quadratic form.

Example 4.9 : *Let A and B be Symmetric n×n matrices whose eigen values are all positive. Show that the eigenvalues of A + B are all positive.*   OR

If A and B are positive definite symmetric $n \times n$ matrices then A + B is also positive definite.

Solution : Given that, A and B are positive definite matrices of order $n \times n$ and are Symmetric.

$\Rightarrow$ All eigen values of A and B both are positive.

Now tr. (A) = sum of all eigen values of A

and tr. (B) = sum of all eigen values of B

$\Rightarrow$ tr. (A) > 0 and tr. (B) > 0 also tr. (A + B) = tr.(A) + tr.(B)

$\Rightarrow$ tr. (A + B) > 0

$\Rightarrow$ rigen values of A + B must be all are positive

$\therefore$ A + B is also positive definite.

Example 4.10 : Let A be $n \times n$ invertible symmetric matrix. Show that if A is positive definite then A^{-1} is also has positive definite quadratic form.

Solution : Step 1 : We know that if $\lambda \neq 0$ is an eigen value of A then $\frac{1}{\lambda} \neq 0$ is an eigenvalue of A^{-1}. It is given that A is $n \times n$ invertible symmetric matrix which is positive definite.

$\therefore$ All the eigen values are positive i.e. $\frac{1}{\lambda_1}$, $\frac{1}{\lambda_2}$, ..., $\frac{1}{\lambda_n}$ all are positive.

Hence, A^{-1} is positive definite.

4.2 Affine Space

As we have seen earlier, given vectors $\vec{v}_1$, $\vec{v}_2$, ..., $\vec{v}_n$ in $\mathbb{R}^n$ one can have linear combination of these vectors as $\alpha_1 \vec{v}_1 + \alpha_2 \vec{v}_2 + ... + \alpha_n \vec{v}_n$ where α_1, α_2, ..., α_n all are scalars.

Definition (Affine Combination) :

Given vectors $\vec{v}_1$, $\vec{v}_2$, ..., $\vec{v}_p$ in $\mathbb{R}^n$ and scalars α_1, α_2, ..., α_p then the linear combination given by,

$\alpha_1 \vec{v}_1 + \alpha_2 \vec{v}_2 + \ .... \ + \alpha_p \vec{v}_p$ is called an affine combination of $\vec{v}_1$, $\vec{v}_2$, ... $\vec{v}_p$ if weights satisfies $\alpha_1 + \alpha_2 + \ .... \ + \alpha_P = 1$

Affine hull or Affine Span : The set of all possible linear combinations of vectors in the set S which satisfies the condition of affine combination is called Affine hull or Affine space

---

Example 4.11 : *Find Affine hull of* $S = \{\vec{v}\}$, $\vec{v} \in \mathbb{R}^n$.

Solution : Given : $S = \{\vec{v}\}$ which is single vector in $\mathbb{R}^n$, then affine combination is $\alpha_1 \vec{v}_1$ with $\alpha_1 = 1$.

$\Rightarrow$ Affine hull of single vector $\{\vec{v}\}$ is again same.

Example 4.12 : *Find Affine hull of* $S = \{\vec{v}_1, \vec{v}_2\}$ *in vector space V.*

Solution :   Given $S = \{\vec{v}_1, \vec{v}_2\}$ then the affine combination of vectors in S is $\vec{w} = \alpha_1 \vec{v}_1 + \alpha_2 \vec{v}_2$, $\vec{w} \in v$ with the condition that

$$\alpha_1 + \alpha_2 = 1$$
$$\Rightarrow \qquad \alpha_1 = 1 - \alpha_2 \qquad\qquad \text{put } \alpha_2 = t \text{ where } t \in \mathbb{R}$$
$$\Rightarrow \qquad \alpha_1 = 1 - t$$

$\therefore$ Affine hull of S has the form, $\vec{w} = (1 - t)\,\vec{v}_1 + t\vec{v}_2$, $t \in \mathbb{R}^n$

Note that, geometrically if $V = \mathbb{R}^2$ then for $S = \{\vec{v}_1, \vec{v}_2\}$ the affine hull of S is Span $\{\vec{v}_1\}$ which is the line through $\vec{v}_1$ (for $t = 0$) parallel to the line passing through origin and $(\vec{v}_2 - \vec{v}_1)$.

Given as follows :

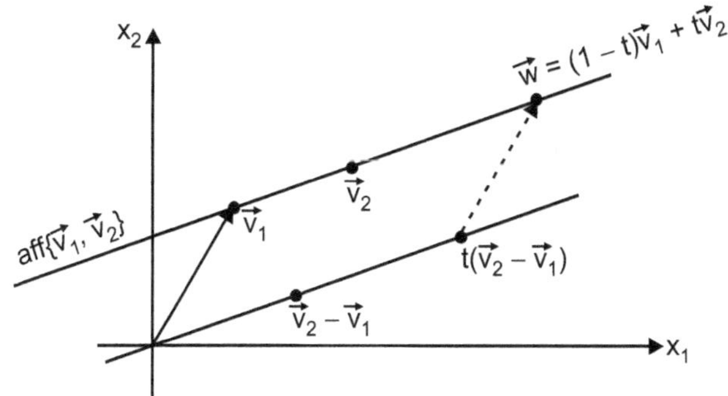

Fig. 4.1

Now, we move towards the necessary and sufficient condition for the vector to be in affine space of given set S.

Theorem 4 : A vector (or print) $\vec{w}$ in $\mathbb{R}^n$ is an affine combination of $\vec{v}_1, \vec{v}_2, ..., \vec{v}_k$ in $\mathbb{R}^n$ if and only if $\vec{w} - \vec{v}_1$ is linear combination of translated vectors (or points) $\vec{v}_2 - \vec{v}_1, \vec{v}_3 - \vec{v}_1, ..., \vec{v}_k - \vec{v}_1$

Proof : Let $S = \{\vec{v}_1, \vec{v}_2 ....., \vec{v}_k\}$ be the set of vectors (or points) in $\mathbb{R}^n$ such that $\vec{w} \in$ affine space of S.

$$\Rightarrow \qquad \vec{w} = \alpha_1 \vec{v}_1 + \alpha_2 \vec{v}_2 + ... + \alpha_k \vec{v}_k \qquad ...(I)$$

with $\qquad \alpha_1 + \alpha_2 + ... + \alpha_k = 1$

$\therefore$ If $\vec{w} - \vec{v}_1$ is in span $\{\vec{v}_2 - \vec{v}_1, \vec{v}_3 - \vec{v}_2, ..., \vec{v}_k - \vec{v}_1\}$

$$\Rightarrow \qquad \vec{w} - \vec{v}_1 = c_2(\vec{v}_2 - \vec{v}_1 + c_3(\vec{v}_3 - \vec{v}_1) + ... + c_k(\vec{v}_k - \vec{v}_1)$$

$$\Rightarrow \qquad \vec{w} - \vec{v}_1 = (1 - c_2 - c_3 ... - c_k) \vec{v}_1 + c_2 \vec{v}_2 + c_3 \vec{v}_3 + ... + c_k \vec{v}_k$$

With the property that, sum scalars on R.H.S. of above equation is 1.

$\therefore$ By definition, $\vec{w} - \vec{v}_1$ is affine combination of $s = \{\vec{v}_1, \vec{v}_2 ...., \vec{v}_k\}$

Step II : Conversely if,

$$\vec{w} = \alpha_1 \vec{v}_1 + \alpha_2 \vec{v}_2 + ... + \alpha_k \vec{v}_k \text{ with } \alpha_1 + \alpha_2 + ... + \alpha_k = 1$$

$$\Rightarrow \qquad \alpha_1 = 1 - \alpha_2 - \alpha_3 ..... - \alpha_k$$

From the above equation we have

$$\vec{w} = (1 - \alpha_2 - \alpha_3 .... - \alpha_k) \vec{v}_1 + \alpha_1 \vec{v}_2 + ... + \alpha_k \vec{v}_k$$

$$\Rightarrow \vec{w} - \vec{v}_1 = \alpha_2(\vec{v}_2 - \vec{v}_1) + \alpha_3(\vec{v}_3 - \vec{v}_1) + ... + \alpha_k(\vec{v}_k - \vec{v}_1)$$

$\therefore$ $\vec{w} - \vec{v}_1$ is in span $\{\vec{v}_2 - \vec{v}_1, \vec{v}_3 - \vec{v}_1, ... \vec{v}_k - \vec{v}_1\}$ $\qquad$ ∎

Example 4.13 : *Write $\vec{y}$ as an affine combination of the vectors given below :*

$$\vec{y} = \begin{bmatrix} 5 \\ 3 \end{bmatrix} \text{ and } S = \left\{ \vec{v}_1 = \begin{bmatrix} 1 \\ 2 \end{bmatrix}, \vec{v}_2 = \begin{bmatrix} -2 \\ 2 \end{bmatrix}, \vec{v}_3 = \begin{bmatrix} 0 \\ 4 \end{bmatrix}, \vec{v}_4 = \begin{bmatrix} 3 \\ 7 \end{bmatrix} \right\}$$

Solution : Step I : By using necessary and sufficient condition for the affine combination.

$$\vec{v}_2 - \vec{v}_1 = \begin{bmatrix} -3 \\ 0 \end{bmatrix}, \ \vec{v}_3 - \vec{v}_1 = \begin{bmatrix} -1 \\ 2 \end{bmatrix}, \ \vec{v}_4 - \vec{v}_1 = \begin{bmatrix} 2 \\ 5 \end{bmatrix} \text{ and } \vec{y} - \vec{v}_1 = \begin{bmatrix} 4 \\ 1 \end{bmatrix}$$

Step II : To find scalars c_2, c_3 and c_4 such that

$$\vec{y} - \vec{v}_1 \;=\; c_2\,(\vec{v}_2 - \vec{v}_1) + c_3\,(\vec{v}_3 - \vec{v}_1) + c_4\,(\vec{v}_4 - \vec{v}_1) \qquad ....(1)$$

Consider the Augmented matrix

$$[A \ b] \;=\; \begin{bmatrix} -3 & -1 & 2 & 4 \\ 0 & 2 & 5 & 1 \end{bmatrix} \sim \begin{bmatrix} 1 & 0 & -3/2 & -3/2 \\ 0 & 1 & 5/2 & 1/2 \end{bmatrix}$$

This shows that, equation (1) is consistent and the general solution is,

$$c_2 = \frac{3}{2}\,c_4 - \frac{3}{2}, \; c_3 = -\frac{5}{2}\,c_4 + \frac{1}{2} \text{ with } c_4 \text{ is free, variable so, put } c_4 = 0$$

$$\Rightarrow \; c_2 \;=\; -\frac{3}{2} \text{ and } c_3 = \frac{1}{2}$$

$$\therefore \; \vec{y} - \vec{v}_1 \;=\; -\frac{3}{2}\,(\vec{v}_2 - \vec{v}_1) + \frac{1}{2}\,(\vec{v}_3 - \vec{v}_1) + 0\,(\vec{v}_4 - \vec{v}_1)$$

$$\therefore \; \vec{y} - \vec{v}_1 \;=\; -\frac{3}{2}\,(\vec{v}_2 - \vec{v}_1) + \frac{1}{2}\,(\vec{v}_3 - \vec{v}_1)$$

$$\Rightarrow \; \vec{y} \;=\; -\frac{3}{2}\,\vec{v}_2 + \frac{3}{2}\,\vec{v}_1 + \frac{1}{2}\,\vec{v}_3 - \frac{1}{2}\,\vec{v}_1 + \vec{v}_1$$

$$\Rightarrow \; \vec{y} \;=\; 2\vec{v}_1 - \frac{3}{2}\,\vec{v}_2 + \frac{1}{2}\,\vec{v}_3$$

with $2 - \dfrac{3}{2} + \dfrac{1}{2} = 1$

$\therefore$ Y is an affine combination of $\vec{v}_1$, $\vec{v}_2$, $\vec{v}_3$ and $\vec{v}_4$.

Example 4.14 : *Write* $\vec{y} = \begin{bmatrix} 17 \\ 1 \\ 5 \end{bmatrix}$ *as affine combination of*

$$S = \left\{ \vec{v}_1 = \begin{bmatrix} -3 \\ 1 \\ 1 \end{bmatrix}, \; \vec{v}_2 = \begin{bmatrix} 0 \\ 4 \\ -2 \end{bmatrix}, \; \vec{v}_3 = \begin{bmatrix} 4 \\ -2 \\ 6 \end{bmatrix} \right\} \; if\ possible.$$

Solution : Step I : $\vec{v}_2 - \vec{v}_1 = \begin{bmatrix} 3 \\ 3 \\ -3 \end{bmatrix}$, $\vec{v}_3 - \vec{v}_1 = \begin{bmatrix} 7 \\ -3 \\ 5 \end{bmatrix}$, $\vec{y} - \vec{v}_1 = \begin{bmatrix} 20 \\ 0 \\ 4 \end{bmatrix}$

Step II : To find c_2, c_3 such that

$$\vec{y} - \vec{v}_1 \;=\; c_2\,(\vec{v}_2 - \vec{v}_1) + c_3\,(\vec{v}_3 - \vec{v}_1) \qquad ...(1)$$

Consider the Augmented matrix

$$[A \ b] \;=\; \begin{bmatrix} 3 & 7 & 20 \\ 3 & -3 & 0 \\ -3 & 5 & 4 \end{bmatrix} \sim \begin{bmatrix} 1 & 0 & 2 \\ 0 & 1 & 2 \\ 0 & 0 & 0 \end{bmatrix}$$

This show that equation (1) is consistent with $c_2 = 2$, $c_3 = 2$

and $\vec{y} - \vec{v}_1 = 2(\vec{v}_2 - \vec{v}_1) + 2(\vec{v}_3 - \vec{v}_1)$

$\Rightarrow \qquad \vec{y} = -3\vec{v}_1 + 2\vec{v}_2 + 2\vec{v}_3$ with $(-3 + 2 + 2 = 1)$

$\therefore$ $\vec{y}$ is an affine combination of $S = \{\vec{v}_1, \vec{v}_2, \vec{v}_3\}$

Note that, if $S = \{\vec{v}_1, \vec{v}_2, \ldots, \vec{v}_n\}$ is itself a basis for $\mathbb{R}^n$ then any vector $\vec{y}$ in $\mathbb{R}^n$ can be written uniquely as linear combination of vectors of S. This combination is affine combination if and only if sum of coefficients of vectors in S is 1.

Example 4.15 : *Let* $S = \left\{ \vec{b}_1 = \begin{bmatrix} 2 \\ 1 \\ 1 \end{bmatrix}, \ \vec{b}_2 = \begin{bmatrix} 1 \\ 0 \\ -2 \end{bmatrix}, \ \vec{b}_3 = \begin{bmatrix} 2 \\ -5 \\ 1 \end{bmatrix} \right\}$ *is an*

orthogonal basis for $\mathbb{R}^3$. *If possible, write* $\vec{y}_1$ *and* $\vec{y}_2$ *as affine combination of vectors in S. Where,*

$$\vec{y}_1 = \begin{bmatrix} 3 \\ 8 \\ 4 \end{bmatrix} \text{ and } \vec{y}_2 = \begin{bmatrix} 6 \\ -3 \\ 3 \end{bmatrix}$$

Solution : Step I : Consider the Augmented matrix [A b] as

$$[A \ \ b] = \begin{bmatrix} 2 & 1 & 2 & 3 & 6 \\ 1 & 0 & -5 & 8 & -3 \\ 1 & -2 & 1 & 4 & 3 \end{bmatrix}$$

$$\uparrow \quad \uparrow \quad \uparrow \quad \uparrow \quad \uparrow$$
$$b_1 \quad b_2 \quad b_3 \quad \vec{y}_1 \quad \vec{y}_2$$

after row reducing the above matrix we get

$$[A \ \ b] \sim \begin{bmatrix} 1 & 0 & 0 & 3 & 2 \\ 0 & 1 & 0 & -1 & 0 \\ 0 & 0 & 1 & -1 & 1 \end{bmatrix}$$

Step II : $\vec{y}_1 = 3\vec{v}_1 - \vec{v}_2 - \vec{v}_3$ and $\vec{y}_2 = 2\vec{v}_1 + 0\vec{v}_2 + \vec{v}_3$

The sum of coefficients in the linear combination for $\vec{y}_1$ is 1 while for $\vec{y}_2$ is 3 which is not 1.

$\therefore$ $\vec{y}_1$ is an affine combination while $\vec{y}_2$ is not, for $S = \{b_1, b_2, b_3\}$

$\boxed{\text{Definition}}$ **(Affine set) :** A set S is affine if for $p, q \in S$

$\Rightarrow (1 - t)p + tq \in S$ for $\forall t \in \mathbb{R}$.

Note that, geometrically Affine set of two vectors is nothing but whenever two vectors in the set then all the vectors on the line passing through these two vectors is nothing but affine set of given set.

$\boxed{\textbf{Theorem 5}}$ A set S is affine if and only if every affine combination of points of S lies in S.

OR

A set S is affine set if and only if S = aff S.

Proof : Part A

Step I : Suppose that S is affine set $\Rightarrow$ if p, q, $\in$ S then

$$(1 - t)\, p + tq \in S \text{ for each real number t.}$$

To show that, S = aff S. we will prove the result by using induction on points of S.

Step I : m = 1 or 2 $\Rightarrow$ S contains one or two points now by the definition of affine set, an affine combination of points of S lies in S.

$\therefore$ aff S $\subseteq$ S is true for m = 1 or 2.

Step II : Assume that result hold for m = k i.e. if S contains k number of points then affine S $\subseteq$ S holds.

Step III : To show that, result hold for m = k + 1. So let, S contains k + 1 points i.e., $S = \left\{ \vec{v_1}, \vec{v_2}, \ldots, \vec{v_k}, \vec{v_{k+1}} \right\}$.

Now as S is affine.

If we consider, $\vec{y} = \alpha_1 \vec{v_1} + \alpha_2 \vec{v_2} + \ldots + \alpha_k \vec{v_k} + \alpha_{k+1} \vec{v_{k+1}}$

Where, $\alpha_1 + \alpha_2 + \ldots + \alpha_k + \alpha_{k+1} = 1$

$\Rightarrow$ atleast one of them must not be equal to 1.

Without loss assume that $\alpha_{k+1} \neq 1$ and define $t = \alpha_1 + \ldots + \alpha_k$

$\Rightarrow t = 1 - \alpha_{k+1} \neq 0$ as $\alpha_{k+1} \neq 1$

$\therefore$ from equation (1) we have

$$\vec{y} = t\left[\frac{c_1}{t}\vec{v_1} + \ldots + \frac{c_k}{t}\vec{k} \right] + c_{k+1} \vec{v_{k+1}}$$

$$\Rightarrow \qquad \vec{y} = (1 - c_{k+1})\left[\frac{c_1}{t}\vec{v_1} + \ldots + \frac{c_k}{t}\vec{v_k} \right] + c_{k+1} \vec{v_{k+1}}$$

Now from **Step II :** $\frac{c_1}{t}\vec{v_1} + \ldots + \frac{c_k}{t}\vec{v_k}$ is in S and $c_{k+1}\vec{v_{k+1}} \in$ S with $(1 - c_{k+1}) + c_{k+1} = 1$

$\Rightarrow \vec{y}$ is in S.

$\therefore$ By the principle of induction every affine combination of two points in $S \Rightarrow \vec{y} \in S$

Hence, aff $S \subseteq S$

Part B : Now to show that $S \subseteq$ aff s which is true as

for $1\vec{v_1} \in S \Rightarrow 1.\vec{v_1} \in S$

$\therefore$ aff $S \subseteq S$

$\therefore$ From Step (A) and Step (B) we have $S =$ aff S

Conversely, if $S =$ aff $S \Rightarrow$ affine combination of two (or more) points of S lies in S.

$\therefore$ S is affine.   ∎

$\boxed{\textbf{Definition}}$: A flat in $\mathbb{R}^n$ is a translate of subspace of $\mathbb{R}^n$. where,

translate of set S in $\mathbb{R}^n$ by a vector $\vec{P}$ is given by

$$S + \vec{P} \;=\; \{\vec{v} + \vec{P} \mid \vec{v} \in S\}$$

Geometrically, translate of set S or subspace S is nothing but translation (or shifting) of every vector in S by some specific vector is given.

Example 4.16 : *Show that if $\vec{P} = \vec{0}$ then for any set S its translate is itself.*

Solution : Given any subset S of $\mathbb{R}^n$ and vector $\vec{P}$ in $\mathbb{R}^n$ then translate of S is given by, $S + \vec{P} = \{\vec{v} + \vec{P} \mid \vec{v} \in S\}$

Now, $\vec{P} = \vec{0}$

$\Rightarrow$ $S + \vec{P} = \{\vec{v} + \vec{0} \mid \vec{v} \in S\}$

                $= \{\vec{v} \mid \vec{v} \in S\}$ $\because \vec{v} + \vec{0} = \vec{v}$ in $\mathbb{R}^n$

                $= S$

$\therefore$ $S + \vec{0} = S$

Example 4.17 : *Show that if S is subspace of $\mathbb{R}^n$ then*

$$S + \overline{P} = S \Leftrightarrow \vec{P} \in S.$$

Solution : We know that, for any set S the translate of S by vector $\in$ $\vec{P}$ in $\mathbb{R}^n$ is given by,

$$S + \vec{P} \;=\; \{\vec{v} + \vec{P} \mid \overline{v} \in S\}$$

$\therefore \qquad S + \overline{P} = S \Leftrightarrow \vec{v} + \vec{P} = \vec{w}$ for $\vec{w} \in S$

$\Leftrightarrow \qquad \vec{P} = \vec{w} - \vec{v} \in S$ - definition of subspace

$\therefore \qquad S + \vec{P} = S \Leftrightarrow \vec{P} \in S$

Note :

(1) Two flats are parallel if one is translate of the other.

(2) The dimension of the flat is the dimension of corresponding parallel subspaces.

(3) The dimension of set S is the dimension of smallest flat containing S. [denoted by dim S]

(4) A line in $\mathbb{R}^n$ is flat of dimension 1.

$\boxed{\textbf{Definition}}$: A hyperplane in $\mathbb{R}^n$ is a flat of dimension 1 less than n that is hyperplane in $\mathbb{R}^n$ is of dimension $n - 1$.

For example :

(1) In $\mathbb{R}^3$ the hyperplanes are planes passing through origin (i.e. $\vec{0}$)

(2) In $\mathbb{R}^2$ the hyperplanes are lines passing through origin (i.e. $\vec{0}$)

(3) The general equation of a hyperplane in $\mathbb{R}^n$ is given by

$$a_1x_1 + a_2x_2 + \ldots + a_nx_n = b$$

Where $a_1, a_2, \ldots, a_n$ and b are constants.

$\boxed{\textbf{Theorem 6}}$: A non-empty set S in $\mathbb{R}^n$ is affine if and only if it is a flat in $\mathbb{R}^n$.

Proof : Step I : Let S be non-empty subset of $\mathbb{R}^n$ such that S is affine.

$\Rightarrow \vec{p}, \vec{q} \in S$ then $(1 - t)\vec{p} + t\vec{q} \in S$ for each real number t.

Let, $\vec{p}$ be any fixed point (vector) in S.

Define $W = S + (-\vec{p}) \Rightarrow S = W + \vec{p}$

To show that, S is flat it is sufficient to show that W is subspace of $\mathbb{R}^n$.

(a) $\vec{p} \in S$ and $W = S + (-\vec{p}) \Rightarrow \vec{0} \in W$

(b) for $\vec{u}_1, \vec{u}_2 \in W \Rightarrow \exists \vec{s}_1$ and $\vec{s}_2 \in S$ such that $\vec{u}_1 = \vec{s}_1 - \vec{p}$

and $\vec{u}_1 = \vec{s}_2 - \vec{p}$

Consider,

$$\vec{u}_1 + t\,\vec{u}_2 = (\vec{s}_1 - \vec{p}) + t(\vec{s}_2 - \vec{p})$$

$$= (1 - t)\vec{s}_1 + t(\vec{s}_1 + \vec{s}_2 - \vec{p}) - \vec{p}$$

As S is affine $\Rightarrow (1-t)\,\vec{s}_1 + t\,(\vec{s}_1 + \vec{s}_2 + \vec{p}\,) \in S$

$\Rightarrow \quad \vec{u}_1 + t\,\vec{u}_2 \in W$

$\therefore$ W is subspace of $\mathbb{R}^n \Rightarrow$ S is flat in $\mathbb{R}^n$

Step II : Conversely suppose, S is flat i.e. S is translate of a subspace (say) W of $\mathbb{R}^n$.

$\Rightarrow S = W + \vec{p}$ for some $\vec{p} \in \mathbb{R}^n$

To show that, S is affine it is equivalent to show that for any pair of vectors $\vec{s}_1,\ \vec{s}_2$ in S then $(1-t)\,\vec{s}_1 + t\vec{s}_2 \in S$.

Now, $\quad S \ = \ W + \vec{p}$ $\qquad\qquad \therefore \vec{w}_1$ and $\vec{w}_2$ in W such that

$$\vec{s}_1 \ = \ \vec{w}_1 + \vec{p}$$

and $\quad \vec{s}_2 \ = \ \vec{w}_2 + \vec{p}$

$\therefore \quad (1-t)\,\vec{s}_1 + t\,\vec{s}_2 = \ (1-t)\,(\vec{w}_1 + \vec{p}\,) + t\,(\vec{w}_2 + \vec{p})$

$$= \ (1-t)\,\vec{w}_1 + t\,\vec{w}_2 + (1-t)\,\vec{p} + t\vec{p}$$

$$= \ (1-t)\,\vec{w}_1 + t\,\vec{w}_2 + \vec{p}$$

$\therefore \quad (1-t)\,\vec{s}_1 + t\,\vec{s}_2 = \ (1-t)\,\vec{w}_1 + t\,\vec{w}_2 + \vec{p} \in W + \vec{P} = S$

As W is subspace of $\mathbb{R}^n \Rightarrow (1-t)\,\vec{w}_1 + t\,\vec{w}_2 \in S$

$(1-t)\,\vec{w}_1 + t\,\vec{w}_2 + \vec{p} \in w + \vec{p} = S$ i.e. $(1-t)\,\vec{s}_1 + t\vec{s}_2 \in S$.

Hence, S is affine. $\qquad\qquad\qquad\qquad\qquad\qquad\qquad\blacksquare$

Note : If $\vec{v}$ in $\mathbb{R}^n$ be any vector then $\tilde{v} = \begin{bmatrix} \vec{v} \\ 1 \end{bmatrix}$ is in $\mathbb{R}^{n+1}$ called as homogenous form of $\vec{v}$

Theorem 7 : A point (or vector) $\vec{y}$ in $\mathbb{R}^n$ is an affine combination of $\vec{v}_1,\ \vec{v}_2,\ \dots\ \vec{v}_p$ in $\mathbb{R}^n$ if and only if the homogeneous form of

$\tilde{y} \in$ span $\{\tilde{v}_1,\ \tilde{v}_2\ \dots,\ \tilde{v}_P\}$ in $\mathbb{R}^n$.

Proof : We know that, $\vec{y}$ is in affine $\{\vec{v}_1,\ \vec{v}_2\dots,\ \vec{v}_P\}$

$\Rightarrow$ there exist weights $\alpha_1,\ \alpha_2,\ \dots\dots,\ \alpha_p$ such that

$\vec{y} = \alpha_1 \vec{v}_1 + \alpha_2 \vec{v}_2 + \dots + \alpha_p \vec{v}_p$ with $\alpha_1 + \alpha_2 + \dots + \alpha_p = 1$

$$\Rightarrow \; \vec{y} = \begin{bmatrix} y \\ 1 \end{bmatrix} = \alpha_1 \begin{bmatrix} \vec{v}_1 \\ 1 \end{bmatrix} + \alpha_2 \begin{bmatrix} \vec{v}_2 \\ 1 \end{bmatrix} + \dots + \alpha_P \begin{bmatrix} \vec{v}_p \\ 1 \end{bmatrix}$$

$$\Rightarrow \; \tilde{y} \in \text{span} \; \{ \tilde{v}_1, \tilde{v}_2, \dots, \tilde{v}_p \} \qquad \blacksquare$$

Remark :

To check whether given vector $\vec{p}$ is an affine combination of $\{ \tilde{v}_1, \tilde{v}_2, \dots, \tilde{v}_k \}$ in $\mathbb{R}^n$. Following steps can be used :

(i)   Write $\tilde{p} = \alpha_1 \tilde{v}_1 + \alpha_2 \tilde{v}_2 + \dots + \alpha_k \tilde{v}_\kappa$

(ii)   Write Augmented matrix $\left[\tilde{v}_1, \tilde{v}_2, \dots, \tilde{v}_\kappa \; \tilde{p} \right]$

(iii) Without any loss, one can write all the entries in the last row (all 1's).

(iv) Row reduced the augmented matrix.

(v)   If it has solution then $\vec{p}$ which can be written as affine combination $\vec{v}_1, \vec{v}_2, \dots, \vec{v}_k$. where, scalar $\alpha_1, \alpha_2, \dots \alpha_p$ are nothing but entries in the last column.

Example 4.18 : *Write* $\vec{y} = \begin{bmatrix} 17 \\ 1 \\ 5 \end{bmatrix}$ *as an affine combination of*

$$\vec{v}_1 = \begin{bmatrix} -3 \\ 1 \\ 1 \end{bmatrix}, \; \vec{v}_2 = \begin{bmatrix} 0 \\ 4 \\ -2 \end{bmatrix}, \; \vec{v}_3 = \begin{bmatrix} 4 \\ -2 \\ 6 \end{bmatrix}, \; \textit{if possible.}$$

Solution : Write $\tilde{y} = \alpha_1 \tilde{v}_1 + \alpha_2 \tilde{v}_2 + \alpha_3 \tilde{v}_3$

$$\Rightarrow \qquad \begin{bmatrix} \vec{y} \\ 1 \end{bmatrix} = \alpha_1 \begin{bmatrix} \vec{v}_1 \\ 1 \end{bmatrix} + \alpha_2 \begin{bmatrix} \vec{v}_2 \\ 1 \end{bmatrix} + \alpha_3 \begin{bmatrix} \vec{v}_3 \\ 1 \end{bmatrix}$$

Step I : The Augmented matrix

$$[\tilde{v}_1 \; \tilde{v}_2 \; \tilde{v}_3 \; \tilde{y}] = \begin{bmatrix} -3 & 0 & 4 & 17 \\ 1 & 4 & -2 & 1 \\ 1 & -2 & 6 & 5 \\ 1 & 1 & 1 & 1 \end{bmatrix}$$

$$\sim \begin{bmatrix} 1 & 1 & 1 & 1 \\ -3 & 0 & 4 & 17 \\ 1 & 4 & -2 & 1 \\ 1 & -2 & 6 & 5 \end{bmatrix} \quad \text{equivalent to 3 row interchanges}$$

$$\sim \begin{bmatrix} 1 & 0 & 0 & -3 \\ 0 & 1 & 0 & 2 \\ 0 & 0 & 1 & 2 \\ 0 & 0 & 0 & 0 \end{bmatrix}$$

$\Rightarrow \alpha_1 = -3, \alpha_2 = 2, \alpha_3 = 2$ with $\alpha_1 + \alpha_2 + \alpha_3 = 1$

and $\vec{y} = \alpha_1 \vec{v_1} + \alpha_2 \vec{v_2} + \alpha_3 \vec{v_3}$

hence, $\vec{y}$ is an affine combination of $\vec{v_1}, \vec{v_2}, \vec{v_3}$

4.3 Affine Dependence and Independence

In the previous semester, we have seen the terminology "linearly dependent" and "linearly independent" of set of vectors. While in previous section, we have compup with affine combination of vectors.

We know that, given set of vectors $S = \{ \vec{v_1}, \vec{v_2}, \vec{v_3} \}$ is said to be linearly dependent if atleast one vector (say $\vec{v}_3$) can be written as linear combination of other vectors i.e.

$$\vec{v_3} = \alpha_1 \vec{v}_1 + \alpha_2 \vec{v}_2$$

Now, if $\alpha_1 = 1 - t, \alpha_2 = t$ for t in $\mathbb{R}$ then

$$\vec{v_3} = (1 - t) \vec{v_1} + t \vec{v_2} \qquad\qquad (\alpha_1 + \alpha_2 = 1)$$

$$\Rightarrow (1 - t) \vec{v_1} + t \vec{v_2} - \vec{v_3} = \vec{0}$$

$\therefore \qquad S = \{ \vec{v_1}, \vec{v_2}, \vec{v_3} \}$ is linearly dependent. Moreover if we put $\alpha_3 = -1$. Then above equation becomes,

$$\alpha_1 \vec{v_1} + \alpha_2 \vec{v_2} + \alpha_3 \vec{v_3} = \vec{0}$$

With the property that $\alpha_1 + \alpha_2 + \alpha_3 = 0$. This gives rise to new defition called affinely dependent and independent.

Definition : Given set $S = \{ \vec{v_1}, \vec{v_2}, \ldots\ldots, \vec{v_k} \}$ in $\mathbb{R}^n$ is said to be affinely dependent if there exist real numbers $\alpha_1, \alpha_2, \ldots. \alpha_k$ not all zero such that

$$\alpha_1 \vec{v_1} + \alpha_2 \vec{v_2} + \ldots\ldots + \alpha_k \vec{v_k} = \vec{0} \text{ and}$$

$$\alpha_1 + \alpha_2 + \ldots.. + \alpha_k = 0. \ \alpha_i \neq 0, \forall i = 1, 2, \ldots., k$$

Note :

(1) If S is not affinely dependent we call it as affinely independent.

(2) If set is Affinely dependent then it is dependent while the converse is not true.

(3) If $S = \{ \vec{v_1} \}$ then it is affinely independent as $\alpha_1 \vec{v_1} = 0 \Rightarrow \alpha_1 = 0$

Example 4.19 : *If* $S = \{\vec{v_1}, \vec{v_2}\}$ *be any set of vectors in* $\mathbb{R}^n$ *then show that S is affinely dependent if and only of* $\vec{v_1} = \vec{v_2}$.

Solution : Let $S = \{\vec{v_1}, \vec{v_2}\}$ be any set of vectors in $\mathbb{R}^n$.

Step I : If S is affinely dependent there exist scalars α_1, α_2 not all zeros such that, $\alpha_1 \vec{v_1} + \alpha_2 \vec{v_2} = \vec{0}$ with $\alpha_1 + \alpha_2 = 0$

$$\Rightarrow \qquad \alpha_1 = -\alpha_2 \neq 0$$

$$\alpha_1 \vec{v_1} + \alpha_2 \vec{v_2} = \vec{0}$$

$$\Rightarrow \qquad \alpha_1 \vec{v_1} = -\alpha_2 \vec{v_2}$$

$$\Rightarrow \qquad \alpha_1 \vec{v_1} = -\alpha_1 \vec{v_2}$$

$$\Rightarrow \qquad \vec{v_1} = \vec{v_2} \qquad\qquad \because \alpha_1 \neq 0$$

Step II : Let $S = \{\vec{v_1}, \vec{v_2}\}$ be such that $\vec{v_1} = \vec{v_2}$ take $\alpha_1 = 1$

and $\alpha_2 = -1$ so that $\alpha_1 \vec{v_1} + \alpha_2 \vec{v_2} = \vec{0}$ with $\alpha_1 + \alpha_2 = 0$

$\therefore$ S is affinely dependent in $\mathbb{R}^n$. ∎

To generalize the above example for more than set of two vectors we have the following theorem.

Theorem 8 : If $S = \{\vec{v_1}, \vec{v_2}, ..., \vec{v_k}\}$ be the given set of vectors in $\mathbb{R}^n$ with $k \geq 2$. Then following statements are equivalent :

(i) S is affinely dependent.

(ii) One of the vector in S is affine combination of other points in S.

(iii) The set $\{\vec{v_2} - \vec{v_1}, \vec{v_3} - \vec{v_1}, ..., \vec{v_k} - \vec{v_1}\}$ in $\mathbb{R}^n$ is linearly dependent.

(iv) The set $\{\tilde{v}_1, \tilde{v}_2, \tilde{v}_3, ..., \tilde{v}_\kappa\}$ of homogeneous forms in $\mathbb{R}^{n+1}$ is linearly dependent.

Proof : Let $S = \{\vec{v_1}, \vec{v_2}, ..., \vec{v_k}\}$ be the set of vectors in $\mathbb{R}^n$ with $k \geq 2$.

Step A : I $\Rightarrow$ II

Let S is affinely dependent $\Rightarrow$ there exists scalars α_1, α_2, ...α_k not all zeros such that $\alpha_1 + \alpha_2 + ... + \alpha_k = 0$ and

$$\alpha_1 \vec{v_1} + \alpha_2 \vec{v_2} + ... + \alpha_k \vec{v_k} = \vec{0} \qquad ...(1)$$

Without loss, assume that $\alpha_1 \neq 0$ so that equation (1) can be rewritten as,

$$\vec{v_1} = \left(-\frac{\alpha_2}{\alpha_1}\right) \vec{v_2} + \left(-\frac{\alpha_3}{\alpha_1}\right) \vec{v_3} + ... + \left(-\frac{\alpha_k}{\alpha_1}\right) \vec{v}_k \quad ...(2)$$

With $1 + \left(\frac{\alpha_2}{\alpha_1}\right) + \left(\frac{\alpha_3}{\alpha_1}\right) + ... + \left(\frac{\alpha_k}{\alpha_1}\right) = 0$

$\therefore$ By definition, $\vec{v_1}$ is affine combination of $\{\vec{v_2}, \vec{v_3}, ..., \vec{v_k}\}$

Step B : II $\Rightarrow$ III

Let $S = \{\vec{v}_1, \vec{v}_2, ..., \vec{v}_k\}$ be the set of vectors in $\mathbb{R}^n$ be such that $\vec{v}_1$ is affine combination of $\{\vec{v}_2, \vec{v}_3, ..., \vec{v}_k\}$

$\Rightarrow \qquad\qquad \vec{v_1} = \alpha_2\vec{v_2} + \alpha_3 \vec{v_3} + ... + \alpha_k\vec{v_k}$

$$\text{with } \alpha_2 + \alpha_3 + ... + \alpha_k = 1$$

Now, $\qquad 1 \cdot \vec{v_1} = \vec{v_1} \qquad\qquad$ ...property of vectors space.

$\therefore$ The above equation becomes,

$$(\alpha_2 + \alpha_3 + ... + \alpha_k) \vec{v_1} = \alpha_2 \vec{v_2} + \alpha_3\vec{v_3} + ... \alpha_k\vec{v_k}$$

$\Rightarrow \alpha_2 (\vec{v_2} - \vec{v_1}) + \alpha_3 (\vec{v_3} - \vec{v_1}) + ... + \alpha_k (\vec{v_k} - \vec{v_1}) = \vec{0}$

and as $\alpha_2 + \alpha_3 + ... + \alpha_k = 1$ not all scalars are zero

so that $\{\vec{v}_2 - \vec{v}_1, \vec{v}_3 - \vec{v}_1, ..., \vec{v}_k - \vec{v}_1\}$ is linearly dependent.

Step C : III $\Rightarrow$ I : If $S = \{\vec{v}_1, \vec{v}_2, ..., \vec{v}_k\}$ is set of vectors in $\mathbb{R}^n$

such that $\{\vec{v}_2 - \vec{v}_1, \vec{v}_3 - \vec{v}_1, ..., \vec{v}_k - \vec{v}_1\}$ in $\mathbb{R}^n$ is linearly dependent.

$\therefore$ there exist α_2, α_3, ..., α_k not all zeros such that

$$\alpha_2 (\vec{v}_2 - \vec{v}_1) + \alpha_3 (\vec{v}_3 - \vec{v}_1) + ... + \alpha_k (\vec{v}_k - \vec{v}_1) = \vec{0}$$

$\Rightarrow \alpha_2 \vec{v_2} + \alpha_3 \vec{v_3} + ... + \alpha_k\vec{v_k} - (\alpha_2 + \alpha_3 + ... + \alpha_k) \vec{v_k} = \vec{0}$

$\Rightarrow (\alpha_2 + \alpha_3 + ... + \alpha_k) \vec{v_1} = \alpha_2 \vec{v_2} + ... + \alpha_k \vec{v_k}$

Define $\qquad\qquad\qquad \alpha_1 = \alpha_2 + \alpha_3 + ... + \alpha_k$

We get, $\qquad\qquad\qquad \alpha_1\vec{v_1} = \alpha_2 \vec{v_2} + ... + \alpha_k\vec{v_k}$

$\Rightarrow -\alpha_1 \vec{v_1} + \alpha_2 \vec{v_2} + \ldots + \alpha_k \vec{v_k} = \vec{0}$ with $-\alpha_1 + \alpha_2 + \ldots + \alpha_k = 0$

Hence, S is affinely dependent.

Step D : IV $\Rightarrow$ (I)

Let $S = \{\vec{v_1}, \vec{v_2}, \ldots, \vec{v_k}\}$ be the set of vectors in $\mathbb{R}^n$ be such that

$\tilde{S} = \{\tilde{v_1}, \tilde{v_2}, \ldots, \tilde{v_k}\}$ in $\mathbb{R}^{n+1}$ is linearly dependent.

$\Rightarrow$ there exist $\alpha_1, \alpha_2, \ldots, \alpha_k$ not all zeros (scalars)

Such that

$$\alpha_1 \tilde{v_1} + \alpha_2 \tilde{v_2} + \ldots + \alpha_k \tilde{v_k} = \tilde{0}$$

$$\Rightarrow \quad \alpha_1 \begin{bmatrix} \vec{v_1} \\ 1 \end{bmatrix} + \alpha_2 \begin{bmatrix} \vec{v_2} \\ 1 \end{bmatrix} + \ldots + \alpha_k \begin{bmatrix} \vec{v_k} \\ 1 \end{bmatrix} = \begin{bmatrix} \vec{0} \\ 0 \end{bmatrix}$$

$$\Rightarrow \quad \begin{bmatrix} \alpha_1 \vec{v_1} \\ \alpha_1 \end{bmatrix} + \begin{bmatrix} \alpha_2 \vec{v_2} \\ \alpha_2 \end{bmatrix} + \ldots + \begin{bmatrix} \alpha_k \vec{v_k} \\ \alpha_k \end{bmatrix} = \begin{bmatrix} \vec{0} \\ 0 \end{bmatrix}$$

$$\Rightarrow \quad \begin{bmatrix} \alpha_1 \vec{v_1} + \alpha_2 \vec{v_2} + \ldots + \alpha_k \vec{v_k} \\ \alpha_1 + \alpha_2 + \ldots + \alpha_k \end{bmatrix} = \begin{bmatrix} \vec{0} \\ 0 \end{bmatrix}$$

We know that, two vectors are same if they are same co-ordinate wise

i.e. $\alpha_1 \vec{v_1} + \alpha_2 \vec{v_2} + \ldots + \alpha_k \vec{v_k} = \vec{0}$ and $\alpha_1 + \alpha_2 + \ldots + \alpha_k = 0$

$\therefore$ By definition, S is affinely dependent. $\blacksquare$

Remark : From above theorem, now one can check whether the given set $S = \{\vec{v_1}, \vec{v_2}, \ldots, \vec{v_k}\}$ in $\mathbb{R}^n$ is affinely dependent or not by finding whether the new set $\{\vec{v_2} - \vec{v_1}, \vec{v_3} - \vec{v_1}, \ldots, \vec{v_k} - \vec{v_1}\}$ is linearly dependent or not.

Example 4.20 : *Check whether*

$S = \left\{\vec{v_1} = \begin{bmatrix} 4 \\ 1 \end{bmatrix}, \vec{v_2} = \begin{bmatrix} 1 \\ 0 \end{bmatrix}, \vec{v_3} = \begin{bmatrix} 5 \\ 4 \end{bmatrix}, \vec{v_4} = \begin{bmatrix} 1 \\ 2 \end{bmatrix}\right\}$ *is affinely dependent or not.*

Solution : Given : $S = \left\{\vec{v_1} = \begin{bmatrix} 4 \\ 1 \end{bmatrix}, \vec{v_2} = \begin{bmatrix} 1 \\ 0 \end{bmatrix}, \vec{v_3} = \begin{bmatrix} 5 \\ 4 \end{bmatrix}, \vec{v_4} = \begin{bmatrix} 1 \\ 2 \end{bmatrix}\right\}$

Consider $S_1 = \left[\vec{v_2} - \vec{v_1} = \begin{bmatrix} -3 \\ -1 \end{bmatrix}, \vec{v_3} - \vec{v_1} = \begin{bmatrix} 1 \\ 3 \end{bmatrix}, \vec{v_4} - \vec{v_1} = \begin{bmatrix} -3 \\ 1 \end{bmatrix}\right]$

and $\alpha_1 (\vec{v_2} - \vec{v_1}) + \alpha_2 (\vec{v_3} - \vec{v_1}) + \alpha_3 (\vec{v_4} - \vec{v_1}) = 0$

$$\Rightarrow \quad \alpha_1 \begin{bmatrix} -3 \\ -1 \end{bmatrix} + \alpha_2 \begin{bmatrix} 1 \\ 3 \end{bmatrix} + \alpha_3 \begin{bmatrix} -3 \\ 1 \end{bmatrix} = \begin{bmatrix} 0 \\ 0 \end{bmatrix}$$

$$\Rightarrow \qquad \begin{bmatrix} -3\alpha_1 + \alpha_2 - 3\alpha_3 \\ -\alpha_1 + 3\alpha_2 + \alpha_3 \end{bmatrix} = \begin{bmatrix} 0 \\ 0 \end{bmatrix}$$

$$\therefore \qquad -3\alpha_1 + \alpha_2 - 3\alpha_3 = 0$$

and
$$-\alpha_1 + 3\alpha_2 + \alpha_3 = 0$$

Put $\alpha_1 = 3\alpha_2 + \alpha_3$ in other equation we get

$$-3(3\alpha_2 + \alpha_3) + \alpha_2 - 3\alpha_3 = 0$$

$$\Rightarrow \qquad -9\alpha_2 - 3\alpha_3 + \alpha_2 - 3\alpha_3 = 0$$

$$\Rightarrow \qquad -8\alpha_2 - 6\alpha_3 = 0$$

$$\Rightarrow \quad \alpha_2 = \frac{3}{4}\alpha_3 \qquad\qquad\qquad \text{(i.e. } \alpha_3 \text{ is free) put } \alpha_3 = 1$$

$$\Rightarrow \quad \alpha_2 = \frac{3}{4} \text{ and } \alpha_1 = \frac{13}{4}$$

$$\therefore \quad \alpha_1 = \frac{13}{14}, \ \alpha_2 = \frac{3}{4} \text{ and } \alpha_3 = 1$$

$\therefore$ S_1 is linearly dependent hence S is affinely dependent.

Example 4.21 : *Check whether*

$$S = \left\{ \vec{v_1} = \begin{bmatrix} 1 \\ 2 \\ -1 \end{bmatrix}, \ \vec{v_2} = \begin{bmatrix} -2 \\ -4 \\ 8 \end{bmatrix}, \ \vec{v_3} = \begin{bmatrix} 2 \\ -1 \\ 11 \end{bmatrix}, \ \vec{v_4} = \begin{bmatrix} 0 \\ 15 \\ -9 \end{bmatrix} \right\} \quad \text{is} \quad \text{affinely}$$

dependent or not.

Solution : Consider,

$$S_1 = \left\{ \vec{v_2} - \vec{v_1} = \begin{bmatrix} -3 \\ -6 \\ 9 \end{bmatrix}, \ \vec{v_3} - \vec{v_1} = \begin{bmatrix} 1 \\ -3 \\ 12 \end{bmatrix}, \ \vec{v_4} - \vec{v_1} = \begin{bmatrix} -1 \\ 13 \\ -8 \end{bmatrix} \right\} \text{ and}$$

$$\alpha_1 (\vec{v_2} - \vec{v_1}) + \alpha_2 (\vec{v_3} - \vec{v_1}) + \alpha_3 (\vec{v_4} - \vec{v_1}) = \vec{0}$$

$$\Rightarrow \qquad \alpha_1 \begin{bmatrix} -3 \\ -6 \\ 9 \end{bmatrix} + \alpha_2 \begin{bmatrix} 1 \\ -3 \\ 12 \end{bmatrix} + \alpha_3 \begin{bmatrix} -1 \\ 13 \\ -8 \end{bmatrix} = \begin{bmatrix} 0 \\ 0 \\ 0 \end{bmatrix}$$

$$\Rightarrow \qquad \begin{bmatrix} -3\alpha_1 + \alpha_2 - \alpha_3 \\ -6\alpha_1 - 3\alpha_2 + 13\alpha_3 \\ 9\alpha_1 + 12\alpha_2 - 8\alpha_3 \end{bmatrix} = \begin{bmatrix} 0 \\ 0 \\ 0 \end{bmatrix}$$

Which gives following equations,

$$-3\alpha_1 + \alpha_2 - \alpha_3 = 0$$

$$-6\alpha_1 - 3\alpha_2 + 13\alpha_3 = 0$$

$$9\alpha_1 + 12\alpha_2 - 8\alpha_3 = 0$$

after solving these equations we get,

$\alpha_1 = 0,\ \alpha_2 = 0,\ \alpha_3 = 0$

$\therefore$ S_1 is linearly independent hence, S is affinely independent.

Example 4.22 : *Show that a set* $\{\vec{v_1},\ \vec{v_2},\ ....,\ \vec{v_k}\}$ *in* $\mathbb{R}^n$ *is affinely dependent when* $k \geq n + 2$.

Solution : Let, $S = \{\vec{v_1},\ \vec{v_2},\ ....,\ \vec{v_k}\}$ be the set of vectors in $\mathbb{R}^n$ with $k \geq n + 2$.

Step I : We know that, every set of vectors containing number of vectors greater than the dimension of $\mathbb{R}^n$ is linearly dependent.

Step II : $S = \{\vec{v_1},\ \vec{v_2},\ ....,\ \vec{v_k}\}$, $k \geq n + 2$ is affinely dependent. If and only if $S_1 = \{\vec{v_2} - \vec{v_1},\ \vec{v_3} - \vec{v_2},\ ....\ \vec{v_k} - \vec{v_1}\}$ contains $k - 1$ number of vectors with $k - 1 \geq n + 1 > n$.

Hence, from step I, S_1 is linearly dependent so that S becomes affinely dependent.

4.4 Convex Combination

In the previous section we have seen how $S = \{\vec{v_1},\ \vec{v_2},\ .....,\ \vec{v_k}\}$ are said to be affinely dependent or independent. Now, we impose more conditions on the scalars $\alpha_1,\ \alpha_2,\ ....,\ \alpha_k$.

Definition : Given set $S = \{\vec{v_1},\ \vec{v_2},\ ....,\ \vec{v_k}\}$ in $\mathbb{R}^n$ then the combination $\alpha_1\vec{v_1} + \alpha_2\vec{v_2} + ... + \alpha_k\vec{v_k}$ is said to be convex combination if

(i) $\alpha_1 + \alpha_2 + ... + \alpha_k = 1$

(ii) $\alpha_j \geq 0,\ j = 1, 2, ...k$

Remark : The set of all convex combinations of vectors (or points) in set S is called the convex hull of S, denoted by ConV S.

Note that, if we see both the definitions i.e. affine S and ConV S then convex hull is properly contained in affine S.

Example 4.23 : *If* $S = \{\vec{v_1}\}$ *in* $\mathbb{R}^n$ *the* $S = conVS$

Solution : Given $S = \{\vec{v_1}\}$ to find $\alpha_1 \in \mathbb{R}$ such that $\alpha_1 = 1$

and $\alpha_i \geq 0$, has the only choice $\alpha_1 = 1$ i.e. $conVS = \{\vec{v_1}\}$

Example 4.24 : *Let*

$$S = \left\{ \vec{v_1} = \begin{bmatrix} 2 \\ 0 \\ -1 \\ 2 \end{bmatrix}, \; \vec{v_2} = \begin{bmatrix} 0 \\ -2 \\ 2 \\ 1 \end{bmatrix}, \; \vec{v_3} = \begin{bmatrix} -2 \\ 1 \\ 0 \\ 2 \end{bmatrix} \right\}$$

Determine whether $\vec{P}$ is in span S, aff S or conV S.

Where, $\vec{P} = \begin{bmatrix} -1 \\ 2 \\ -3/2 \\ 5/2 \end{bmatrix}$

Solution : Given $S = \{ \vec{v_1}, \vec{v_2}, \vec{v_3} \}$, and $\vec{p} = \begin{bmatrix} -1 \\ 2 \\ -3/2 \\ 5/2 \end{bmatrix}$

(I) Let W be the subspace of $\mathbb{R}^4$ spanned by S, where S is an orthogonal set.

$$\therefore \qquad \vec{v_1} \cdot \vec{v_2} = 0 = \vec{v_2} \cdot \vec{v_3} = \vec{v_1} \cdot \vec{v_3}$$

To show that, $\vec{P}$ is in Span S it is equivalent to show that,

$\text{Proj}_W \vec{P} = \vec{P}$

Now, $\text{Proj}_W \vec{P} = \dfrac{\vec{P} \cdot \vec{v_1}}{\vec{v_1} \cdot \vec{v_1}} \vec{v_1} + \dfrac{\vec{P} \cdot \vec{v_2}}{\vec{v_2} \cdot \vec{v_2}} \vec{v_2} + \dfrac{\vec{P} \cdot \vec{v_3}}{\vec{v_3} \cdot \vec{v_3}} \vec{v_3}$

$$= \frac{9/2}{9} \vec{v_1} + \frac{7/2}{9} \vec{v_2} + \frac{9}{9} \vec{v_3}$$

$$= \frac{1}{2} \vec{v_1} + \frac{7}{18} \vec{v_2} + \vec{v_3}$$

$$= \begin{bmatrix} -1 \\ 2/9 \\ 5/18 \\ 4 \end{bmatrix} \neq \vec{P}$$

$\therefore$ $\vec{P}$ is not in Span S.

(II) As $\alpha_1 = \dfrac{1}{2}$, $\alpha_2 = \dfrac{7}{18}$, $\alpha_3 = 1 \Rightarrow \alpha_1 + \alpha_2 + \alpha_3 \neq 1$

$\therefore$ $\vec{P} \notin$ aff S and hence $\vec{P}$ is not in conV S

Example 4.25 : *Let* $S = \left\{ \vec{v_1} = \begin{bmatrix} 2 \\ 0 \\ 0 \end{bmatrix}, \vec{v_2} = \begin{bmatrix} 0 \\ 2 \\ 0 \end{bmatrix}, \vec{v_3} = \begin{bmatrix} 0 \\ 0 \\ 2 \end{bmatrix} \right\}$ *and*

$\vec{P} = \begin{bmatrix} 1 \\ 3 \\ 1 \end{bmatrix}$. *Determine whether* $\vec{P}$ *is in Span S, aff S or ConV S.*

Solution : Given, $S = \{ \vec{v_1}, \vec{v_2}, \vec{v_3} \}$ and $\vec{P} = \begin{bmatrix} 1 \\ 3 \\ 1 \end{bmatrix}$

(I) As S is an orthogonal set. Let W be the subspace spanned by S. To check whether $\vec{P}$ is in Span S it is equivalent to show that,

$$\vec{P} = \text{Proj}_w \, \vec{P}$$

$$\text{Now, Proj}_w \, \vec{P} = \frac{\vec{P} \cdot \vec{v_1}}{\vec{v_1} \cdot \vec{v_1}} \vec{v_1} + \frac{\vec{P} \cdot \vec{v_2}}{\vec{v_2} \cdot \vec{v_2}} \vec{v_2} + \frac{\vec{P} \cdot \vec{v_3}}{\vec{v_3} \cdot \vec{v_3}} \vec{v_3}$$

$$= \frac{2}{4} \vec{v_1} + \frac{6}{4} \vec{v_2} + \frac{2}{4} \vec{v_3} = \vec{P}$$

$\therefore$ $\vec{P}$ is in Span S.

(II) Now, $\vec{P} = \frac{1}{2} \vec{v_1} + \frac{3}{2} \vec{v_2} + \frac{1}{2} \vec{v_3}$

$\therefore$ $\alpha_1 = \frac{1}{2}, \alpha_2 = \frac{3}{2}, \alpha_3 = \frac{1}{2}$ with $\alpha_1 + \alpha_2 + \alpha_3 = \frac{1}{2} + \frac{3}{2} + \frac{1}{2} \neq 1$

$\therefore$ $\vec{P}$ is not in aff S hence $\vec{P}$ is not in ConV S.

Definition : A set S convex if for each $\vec{p}$, $\vec{q} \in$ S the line segment $\vec{pq}$ is in S.

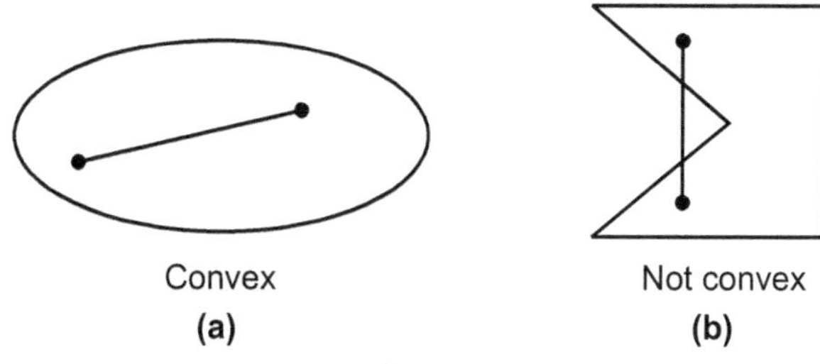

Fig. 4.2

Following theorem will give necessary as well as sufficient condition for the Set S to become convex. This is because one cannot check each and every point of S.

If S is finite then it is easy to check but if S is infinite set then it is quite difficult to check convex property.

Theorem 9 : A set S is convex if and only if every convex combination of points of S lies in S i.e. $S = \text{conV } S$

Step I : Every point (or vector) in S say $\vec{v}$ can be written as

$$\alpha_1 \vec{v_1} = 1\vec{v_1} \text{ i.e. } \alpha_1 = 1 > 0.$$

Condition holds for the single vector of ConV S

Hence, every vector in S is in ConV S.

Step II : Conversely to show that, every convex combination of such points (or vectors) lies in S.

We will prove the statement by using mathematical induction.

(a) From step (I) the result hold for $k = 1$.

(b) Assume that the result hold for $1 \leq j \leq k$

(c) Consider, convex combination of $k + 1$ points (or vectors)

$$\text{i.e. } \vec{v} = \alpha_1 \vec{v}_1 + \alpha_2 \vec{v}_2 + \ldots + \alpha_k \vec{v}_k + \alpha_{k+1} \vec{v}_{k+1} \qquad \ldots(1)$$

with $\alpha_1 + \alpha_2 + \ldots + \alpha_k + \alpha_{k+1} = 1$ and $0 \leq \alpha_i \leq 1$

for $i = 1, 2, \ldots, k+1$

If $\alpha_{k+1} = 1$ then $\alpha_1 = \alpha_2 \ldots = \alpha_k = 0$ and $\vec{y} = \vec{v}_{k+1} \in S$

If $\alpha_{k+1} < 1$ consider, $t = 1 - \alpha_{k+1}$ where $t = \alpha_1 + \alpha_2 + \ldots + \alpha_k$

$\Rightarrow t = 1 - \alpha_{k+1} > 0$ as $\alpha_{k+1} < 1$

Then from equation (1)

$$\vec{y} = (1 - c_{k+1})\left[\frac{\alpha_1}{t}\vec{v}_1 + \frac{\alpha_2}{t}\vec{v}_2 + \ldots + \frac{\alpha_k}{t}\vec{v}_k\right] + \alpha_{k+1}\vec{v}_{k+1}$$

with $\quad \dfrac{\alpha_1}{t} + \dfrac{\alpha_2}{t} + \ldots \dfrac{\alpha_k}{t} = 1 \qquad\qquad \because \alpha_1 + \alpha_2 + \ldots + \alpha_k = t$

Now, define $\quad \vec{w} = \dfrac{\alpha_1}{t}\vec{v} + \dfrac{\alpha_2}{t}\vec{v}_2 + \ldots + \dfrac{\alpha_k}{t}\vec{v}_k \in S$

$\Rightarrow \qquad\qquad \vec{y} = (1 - \alpha_{k+1})\vec{w} + \alpha_{k+1}\vec{v}_{k+1}$

$$\text{with } (1 - \alpha_{k+1}) + \alpha_{k+1} = 1$$

$\therefore$ $\vec{y}$ is in S $\Rightarrow$ Every convex combination of points lies in S.

Hence, S = ConV S.   ■

Remark : If S_1 is positive quadrant in $\mathbb{R}^2$ and S_2 is negative quadrant in $\mathbb{R}^2$ then both are convex while if we consider $S_1 \cup S_2$ then it is not convex as $S_1 \cup S_2$ does not contain X-axis. So that, the line from any point from S_1 to point in S_2 does not contained in $S_1 \cup S_2$ completely.

The result for intersection holds i.e. S_1 and S_2 are convex then $S_1 \cap S_2$ is again convex. This result can be generalize as follows :

$\boxed{\textbf{Thorem 10}}$ If $\{S_\alpha : \alpha \in \Omega \}$ be any arbitrary collection of convex sets. Then $\underset{\alpha\in\Omega}{\cap} S_\alpha$ is convex.

Proof : Given collection of arbitrary convex sets $\{S_\alpha : \alpha\in\Omega\}$ where, Ω is any indexed set.

To show that, $\underset{\alpha\in\Omega}{\cap} S_\alpha$ is again convex.

Let $\vec{u}$, $\vec{w}$ $\underset{\alpha\in\Omega}{\cap} S_\alpha \Rightarrow \vec{u}$, $\vec{w} \in S_\alpha$ for each $\alpha \in \Omega$

As each S_α is convex the line segment joining there two vectors (points) $\vec{u}$ and $\vec{w}$ i.e. $\vec{uw}$ is in S_α for every $\alpha\in \Omega$

$\Rightarrow \vec{uw} \in \underset{\alpha\in\Omega}{\cap} S_\alpha$

$\therefore \underset{\alpha\in\Omega}{\cap} S_\alpha$ is again convex.   ■

Remark :

1. The above result can be extend in similar manner. For affine S. i.e. arbitrary intersection of affine sets is again affine.

2. We know that, if $\vec{p}$ is in ConV S then by definition, $\vec{p}$ can be written as convex combination of points (or vectors) in S satisfying the condition. It was the, caratheodory a Greek mathematician who have given the result for the 'number of vectors (or points) required to write this convex combination" in 1907, for $\mathbb{R}^n$.

To prove the result, we will use the example 4.22.

Theorem 11 (Carathodory Theorem) :

If S is non-empty subset of $\mathbb{R}^n$, then every point (or vector) in conV S can be expressed as a convex combination of (n + 1) or fewer points (or vectors) of S.

Proof : Let $\vec{p}$ be any point (or vector) in convex hull of S. Then $\vec{p}$ is a convex combination of finite number of points (or vectors) in S.

i.e. $S \subset \mathbb{R}^n$ and

$$\vec{p} \;=\; \alpha_1 \vec{v}_1 + \alpha_2 \vec{v}_2 + \ldots + \alpha_k \vec{v}_k \qquad \ldots(1)$$

where, every $\vec{v}_j \in S$ and $\alpha_j \geq 0, j = 1, 2, \ldots, k$

To show that, $k \leq n + 1$. If $k > n + 1 \Rightarrow k \geq n + 2$.

Now, we know that, every set in $\mathbb{R}^n$ having atleast $(n + 2)$ number of points (or vectors) is affinely dependent.

$\Rightarrow$ There exist scalars $\beta_1, \beta_2, \ldots, \beta_k$ not all zeros such that

$$\beta_1 \vec{v}_1 + \beta_2 \vec{v}_2 + \ldots + \beta_k \vec{v}_k \;=\; \vec{0} \qquad \ldots(2)$$

With $$\beta_1 + \beta_2 + \ldots + \beta_k \;=\; 0 \qquad \ldots(A)$$

Solving equation (1) and (2) simultaneously and eliminate $\vec{v}_I$ we obtain convex combination of $(k - 1)$ vectors of S which is equal to $\vec{p}$

Without any loss assume that $\beta_k \neq 0$, i.e. $\beta_k > 0$ $\qquad (\because \beta_j \geq 0, \forall j)$

Such that, $\dfrac{\alpha_k}{\beta_k} \leq \dfrac{\alpha_i}{\beta_i}$ for those i for which $\beta_i > 0, i = 1, 2, \ldots, k$

Define $$b_j \;=\; \alpha_j - \left(\frac{\alpha_k}{\beta_k}\right) \beta_j \qquad \text{then } b_k = 0$$

And $$\sum_{j=1}^{k} b_j \;=\; \sum_{j=1}^{k} \alpha_j - \frac{\alpha_k}{\beta_k} \sum_{j=1}^{k} \beta_j$$

$$=\; 1 - \frac{\alpha_k}{\beta_k} \, 0 = 1$$

From (A)

Also, $b_j \geq 0$ for every $j = 1, 2, \ldots, k$

Such that

$$b_1 \vec{v}_1 + b_2 \vec{v}_2 + \ldots b_{k-1}\vec{v}_{k-1} = b_1\vec{v}_1 + \ldots + b_{k-1}\vec{v}_{k-1} + b_k\vec{v}_k$$

$$\therefore (b_k = 0)$$

$$= \left[\alpha_1 - \frac{c_k}{\beta_k}\beta_1\right]\vec{v}_1 + \ldots + \left[\alpha_k - \frac{c_k}{\beta_k}\cdot\beta_\kappa\right]\vec{v}_k$$

$$= \sum_{j=1}^{k}\alpha_j\vec{v}_j + \frac{c_k}{\beta_k}\sum_{j=1}^{k}\beta_j\vec{v}_j$$

$$\Rightarrow \sum_{j=1}^{k-1}b_j\vec{v}_j = \sum_{j=1}^{k-1}\alpha_j\vec{v}_j = \bar{p} \qquad \left[\because \sum_{j=1}^{k}\beta_j\vec{v}_j = 0 \text{ from (2)}\right]$$

i.e. $\vec{p}$ is linear combination of $(k-1)$ vectors in S with

$$b_1 + b_2 + \ldots + b_{k-1} = 1$$

and
$$b_j \geq 0, \forall j$$

$\Rightarrow \vec{p}$ becomes convex combination of $(k-1)$ vectors

where, $(k-1) \geq n+1$ if $(k-1) > n+1$ repeat the above process until $\vec{p}$ can be expressed as convex combination of atmost $(n+1)$ number of points (or vectors) in S. ∎

Example 4.26 : By using Caratheodary's theorem, if possible express $\vec{p}$ as combination of three points of S where,

$$S = \left\{\vec{v}_1 = \begin{bmatrix}2\\0\end{bmatrix}, \vec{v}_2 = \begin{bmatrix}4\\6\end{bmatrix}, \vec{v}_3 = \begin{bmatrix}10\\8\end{bmatrix}, \vec{v}_4 = \begin{bmatrix}6\\0\end{bmatrix}\right\} \text{ and } \vec{p} = \begin{bmatrix}20/3\\5\end{bmatrix}$$

Solution : Given : $S = \{\vec{v}_1, \vec{v}_2, \vec{v}_3, \vec{v}_4\}$ and $\vec{p} = \begin{bmatrix}20/3\\5\end{bmatrix}$

Then $\vec{p}$ can be written as linear combination of vectors in S as,

$$\frac{1}{2}\vec{v}_1 + \frac{1}{3}\vec{v}_2 + \vec{v}_3 + \frac{1}{6}\vec{v}_4 = \vec{p} \qquad \ldots(1)$$

The set S is affinely dependent, as we have

$$\alpha_1\vec{v}_1 + \alpha_2\vec{v}_2 + \alpha_3\vec{v}_3 + \alpha_4\vec{v}_4 = \vec{0} \qquad \ldots(2)$$

Where, $\alpha_1 = -10, \alpha_2 = 8, \alpha_3 = -6, \alpha_4 = 8$

Choose $\vec{v}_2$ and $\vec{v}_4$ in equation (2) whose coefficients are positive.

For each vector, compute the ratio of the coefficients in equation (1) and (2). The ratio for $\vec{v_2}$ is $\dfrac{1}{3} \div 8 = \dfrac{1}{24}$ and that for $\vec{v_4}$ is $\dfrac{1}{6} \div 8 = \dfrac{1}{48}$

Now, the ratio, $\dfrac{1}{48} < \dfrac{1}{24}$

$\therefore$ Subtract $\dfrac{1}{48}$ equation (2) from equation (1) i.e. (1) $- \dfrac{1}{48}$ equation (2) to eliminate $\vec{v_4}$.

$$\left(\frac{1}{2} + \frac{10}{48}\right) \vec{v_1} + \left(\frac{1}{3} - \frac{8}{48}\right) \vec{v_2} + \left(1 + \frac{6}{48}\right) \vec{v_3} + \left(\frac{1}{6} - \frac{8}{48}\right) \vec{v_4} = \vec{p}$$

$$\Rightarrow \frac{34}{28} \vec{v_1} + \frac{8}{48} \vec{v_2} + \frac{54}{48} \vec{v_3} = \vec{p}$$

Think Over It

1. Can you find some applications of Quadratic forms in coding Theory.
2. Explain how convex hull were used to avoid car Collision ?

Summary

1. Quadratic Forms are the polynomials of degree 2.

2. If A is Symmetric then the coefficient of $x_i \, x_j \, i \neq j$ is always even number in Quadratic form.

3. If A is diagonalizable matrix then by change of variable the new quadratic form does not contain any Cross-product term.

4. $Q(X)$ is mainly divided into positive, negative and indefinite forms based on value.

5. If $\vec{u} \in$ affine S then $\vec{u} \in$ span S but not conversely.

6. Geometrically affine set of two vectors is nothing by line passing through these two vectors.

7. A flat in $\mathbb{R}^n$ is a translate of a subspace of $\mathbb{R}^n$.

8. Every affinely dependent set is linearly dependent but not conversely.

9. Every set S having more that $(n + 1)$ vectors of $\mathbb{R}^n$ is always affinely dependent.

10. If $\vec{u} \in$ conv. $S \Rightarrow \vec{u} \in$ affs $\Rightarrow \vec{u} \in$ span S but not conversely.

Exercise

[A] Say True or False : Justify !

1. If A is symmetric matrix the cross product term coefficient in Quadratic form is always odd.

2. If A is non-symmetric then cross product term coefficients in Quadratic form is always zero.

3. If A is Non-Symmetric matrix then its Quadratic form is given by XA^tX.

4. Quadratic form of 2×2 matrix A has more than 4 terms.

5. The Quadratic form of real symmetric matrix A of order $n \times n$ is always unique.

6. If $Q(X) = Y^t DY$ is Quadratic form of matrix A obtained by change of variables then D is always invertible.

7. $Q(X)$ is positive definite if and only if the eigenvalues of A are all positive.

8. If A and B are positive definite then $A + B$ is also positive definite.

9. If A is $n \times n$ invertible symmetric matrix and A is positive definite then A^{-1} is also positive definite

10. Every Affinely dependent set is linearly dependent.

11. A Flat is a dialiation of subspace of $\mathbb{R}^n$.

12. Set of two vectors is affinely dependent if and only if they are independent.

[B] Multiple Choice Questions : Choose the Correct Alternative :

1. If A is 2×2 matrix having Quadratic form $Q(X) = a_{11} x_1^2 + a_{12}x_1x_2 + a_{21}x_2x_1 + a_{22} x_2^2$ then for symmetric matrix; ....

   (a) a_{12}, a_{21} both are odd   (b) a_{12}, a_{21} both are even

   (c) a_{12}, a_{21} both are prime   (d) None of these

2. If A is 2×2 Symmetric matrix and diagonal elements are zero then its quadratic form were ....

(a) $a_{12} x_1 x_2 + a_{21} x_2 x_1$

(b) $x_1^2 + x_2^2$

(c) $x_1 x_1^2 + x_{12} + x_{21} + x_2^2$

(d) None of these

3. If A is Symmetric then its quadratic form $Q(X)$ ....

(a) $X\,AX$

(b) $X^t\,AX$

(c) $X^t\,A^t\,X^t$

(d) None of these

4. If $Q(X) = 3x_1^2 - 4x_1 x_2 + 7x_2^2$ is quadratic form of symmetric matrix A then A is ....

(a) $A = \begin{bmatrix} 0 & 4 \\ 3 & 0 \end{bmatrix}$

(b) $A = \begin{bmatrix} 3 & 2 \\ 2 & 7 \end{bmatrix}$

(c) $A = \begin{bmatrix} 3 & -2 \\ -2 & 7 \end{bmatrix}$

(d) None of these

5. If $A \begin{bmatrix} 3 & 2 & 0 \\ 2 & 2 & 2 \\ 0 & 2 & 1 \end{bmatrix}$ then A is; ....

(a) Positive definite

(b) Negative definite

(c) indefinite

(d) None of these

6. If A is $n \times n$ invertible symmetric matrix which is positive definite the A^{-1} is ......

(a) Positive definite

(b) Negative definite

(c) indefinite

(d) None

7. If $\alpha_1 \vec{v_1} + \alpha_2 \vec{v_2} + ... + \alpha_n \vec{v_n}$ is an affine combination of $\vec{v_1}, \vec{v_2}, ...$ $\vec{v_n}$ then ....

(a) $\alpha_1 + \alpha_2 + ... + \alpha_n > 1$

(b) $\alpha_1 + \alpha_2 + ... + \alpha_n = 0$

(c) $\alpha_1 + \alpha_2 + ... + \alpha_n = 1$

(d) $\alpha_1 + \alpha_2 + ... + \alpha_n < 1$

8. If S is subset of $\mathbb{R}^n$ which is affinely dependent then S is ....

(a) S is linearly independent

(b) S is linearly dependent

(c) S is Basis

(d) None of these

9. The Affine hull of single vector is ....

(a) $\mathbb{R}^n$

(b) empty set

(c) Set itself

(d) None

10. The hyper planes in $\mathbb{R}^3$ are ....

    (a) lines in $\mathbb{R}^2$

    (b) planes passing through origin in $\mathbb{R}^3$

    (c) X – Axis   (d) None of these

[C] Theory Questions :

1. Show that, if A Symmetric $n \times n$ real matrix then A and A^t have same quadratic forms.

2. Show that if A is $n \times n$ positive definite matrix then there exist a positive definite matrix B such that $A = B^t B$.

3. If $A = \begin{bmatrix} a & b \\ c & d \end{bmatrix}$ Show that Q(X) (quadratic form of A) is positive definite if $\det A > 0$ and $a > 0$

4. Show that Quadratic form of real symmetric matrix is unique.

5. If A is positive definite matrix with eigen values $\lambda_1, \lambda_2, ..., \lambda_n$ all the positive then show that A^{-1} is also positive definite having rigen values $\dfrac{1}{\lambda_1}, \dfrac{1}{\lambda_2}, ..., \dfrac{1}{\lambda_n}$.

6. If S is affinely dependent subset of $\mathbb{R}^n$ then show that S is linearly dependent.

7. Show that set S is affine if and only if Every combination of points of S lies in S.

8. Show that set $S = \{ \vec{v_1}, \vec{v_2}, ... \vec{v_k} \}$ in $\mathbb{R}^n$ is affinely dependent $k \geq n + 2$.

9. Explain with an example that, "Union of two convex sets need not be convex".

10. Show that $S = \{\vec{v_1}, \vec{v_2}\}$ in $\mathbb{R}^n$ is affinely dependent if and only if $\vec{v_1} = \vec{v_2}$.

11. Show that for any set S in $\mathbb{R}^n$, the convex hull of S is the intersection of all the convex sets that contain S.

[D] Numerical Problems :

1. If $Q(X) = 5x_1^2 + 3x_2^2 + 2x_3^2 - x_1x_2 + 8x_2x_3$. Find symmetric matrix A whose quadratic equation is Q(X) and write it in the form $X^t AX$.

2. If $Q(X) = x_1^2 - 8x_1x_2 - 5x_2^2$. Compute the value of Q(X) for $X = \begin{bmatrix} -3 \\ 1 \end{bmatrix}, \begin{bmatrix} -2 \\ 2 \end{bmatrix}$.

3. Write in the form of X^tAX for the quadratic form

$Q(X) = x_1^2 - 8x_1x_2 - 5x_2^2$ and then by making change of variable, find Quadratic form which does not contain cross product term.

4. Is $Q(X) = 3x_1^2 + 2x_2^2 + x_3^2 + 4x_1x_2 + 4x_2x_3$ positive definite ?

5. Describe a positive semidefinite matrix A in term of its eigenvalues.

6. If $S \{ \vec{v_1}, \vec{v_2}, \vec{v_3}\ \vec{v_4}\}$ and $\vec{y} = \begin{bmatrix} 4 \\ 2 \end{bmatrix}$ write $\vec{y}$ as an affine combination of

$\vec{v_1}, \vec{v_2}, \vec{v_3}, \vec{v_4}$. Where, $\vec{v_1} = \begin{bmatrix} 1 \\ 2 \end{bmatrix}, \vec{v_2} = \begin{bmatrix} 2 \\ 5 \end{bmatrix}, \vec{v_3} = \begin{bmatrix} 1 \\ 3 \end{bmatrix}, \vec{v_4} = \begin{bmatrix} -2 \\ 2 \end{bmatrix}$.

7. Determine whether $\vec{u} = \begin{bmatrix} 2 \\ 0 \\ 0 \end{bmatrix}$ and $\vec{v} = \begin{bmatrix} 1 \\ 2 \\ 2 \end{bmatrix}$ are affine

combination of $S = \left\{ \vec{b_1} = \begin{bmatrix} 4 \\ 0 \\ 3 \end{bmatrix}, \vec{b_2} = \begin{bmatrix} 0 \\ 4 \\ 2 \end{bmatrix}, \vec{b_3} = \begin{bmatrix} 5 \\ 2 \\ 4 \end{bmatrix} \right\}$

8. Determine whether $S = \{\vec{v_1}, \vec{v_2}, \vec{v_3}\}$ is affinely independent, where

$\vec{v_1} = \begin{bmatrix} 1 \\ 3 \\ 7 \end{bmatrix}, \vec{v_2} = \begin{bmatrix} 2 \\ 7 \\ 13/2 \end{bmatrix}, \vec{v_3} = \begin{bmatrix} 0 \\ 4 \\ 7 \end{bmatrix}$

9. Let $S = \{\vec{v_1}, \vec{v_2}, \vec{v_3}\}$ and $\vec{P} = \begin{bmatrix} 0 \\ 3 \\ 3 \\ 0 \end{bmatrix}$ Determine whether $\vec{P}$ is in span

S, aff A and conV S. where, $\vec{v_1} = \begin{bmatrix} 3 \\ 0 \\ 6 \\ -3 \end{bmatrix}, \vec{v_2} = \begin{bmatrix} -6 \\ 3 \\ 3 \\ 0 \end{bmatrix}, \vec{v_3} = \begin{bmatrix} 3 \\ 6 \\ 0 \\ 3 \end{bmatrix}$

10. By using Caratherodory's theorem express $\vec{u}$ as convex combination of vectors of S. Where,

$S = \left\{ \vec{v_1} = \begin{bmatrix} 1 \\ 0 \end{bmatrix}, \vec{v_2} = \begin{bmatrix} 2 \\ 3 \end{bmatrix}, \vec{v_3} = \begin{bmatrix} 5 \\ 4 \end{bmatrix}, \vec{v_4} = \begin{bmatrix} 3 \\ 0 \end{bmatrix} \right\}$.

Answers

[A] (1) False (2) False (3) False (4) False (5) True (6) False (7) True

(8) True (9) True (10) True (11) False (12) False.

[B] (1) - (b)   (2) - (a)   (3) - (b)   (4) - (c)   (5) - (c)   (6) - (a)   (7) - (c)   (8) - (b)   (9) - (c)   (10) - (b).

[D] 1. $A = \begin{bmatrix} 5 & -1/2 & 0 \\ -1/2 & 3 & 4 \\ 0 & 4 & 2 \end{bmatrix}$, $Q(X) = [x_1\ x_2\ x_3]\ A \begin{bmatrix} x_1 \\ x_2 \\ x_3 \end{bmatrix}$

2. $Q(-3, 1) = 28$, $Q(-2, 2) = 16$

3. $A = \begin{bmatrix} 1 & -1 \\ -4 & -5 \end{bmatrix}$, $Q(X) = 3y_1^2 - 7y_2^2$ by making change of variable $x = Py$.

4. $Q(X)$ is indefinite not positive definite.

5.   All rigen values must be non-negative.

6. $\vec{y} = 13\vec{v_1} + 6\vec{v_2} - 19\vec{v_3} + \vec{v_4}$

7. $\vec{u} = -2\vec{b_1} - \vec{b_2} + 2\vec{b_3}$ is not affine Combination as $-2 - 1 + 2 \neq 1$

$\vec{v} = \dfrac{2}{3}\vec{b_1} + \dfrac{2}{3}\vec{b_2} - \dfrac{1}{3}\vec{b_3}$ is affine Combination as $\dfrac{2}{3} + \dfrac{2}{3} - \dfrac{1}{3} = 1$

8.   Yes, S is affinely independent.

9. $\vec{P} = \dfrac{1}{3}\vec{v_1} + \dfrac{1}{3}\vec{v_2} + \dfrac{1}{3}\vec{v_3}$

$\therefore\ \vec{P} \in$ span S, $\dfrac{1}{3} + \dfrac{1}{3} + \dfrac{1}{3} = 1$ $\therefore\ \vec{P} \in$ aff S moreover,

all are non-negative hence $\vec{P} \in$ conV S.

10. $\vec{\mu} = \dfrac{17}{48}\vec{v_1} + \dfrac{4}{48}\vec{v_2} + \dfrac{27}{48}\vec{v_3}$

Appendix

(A) Vector Spaces

Q.1. Show that following sets are vector spaces over the field of real numbers $\mathbb{R}$.

(a) $V = \{a_0 + a_1 x + a_2 x^2 + \ldots + a_0 x^n \mid a_i \in \mathbb{R}, 1 \leq i \leq n\}$ set of all the polynomials of degree 'n' under the addition and scalar multiplication defined as follows :

$$p(x) + q(x) = (a_0 + b_0) + (a_1 + b_1) x + \ldots + (a_n + b_n) x^n \text{ and}$$

$$\alpha p(x) = \alpha a_0 + \alpha a_1 x + \alpha a_2 x^2 + \ldots + \alpha a_n x^n, \ \alpha \in \mathbb{R},$$

where, $p(x), q(x) \in V$.

(b) $V = \{[a_{ij}]_{m \times n} \mid a_{ij} \in \mathbb{R}\}$ set of all matrices of order $m \times n$ with real entries under usual matrix addition and matrix scalar multiplication over $\mathbb{R}$.

(c) Let V be the set of all functions from $\mathbb{R}$ to $\mathbb{R}$, with addition and scalar multiplication defined as :

$$(f + g)(x) = f(x) + g(x) \text{ and } (\alpha f)(x) = \alpha f(x) \ \forall \ x \in \mathbb{R}.$$

Show that V is vector space over $\mathbb{R}$.

Q.2. Show that following sets are not vector space. Give suitable reason.

(a) $V = \{(a, b) \mid a, b \in \mathbb{R}\}$ under addition and scalar multiplication defined as :

$$(a, b) + (c, d) = (a + c, b + d) \text{ and } \alpha(a, b) = (\alpha a, b) \text{ for } \alpha \in \mathbb{R}.$$

(b) $V = \{(a, b) \mid a, b \in \mathbb{R}\}$ under addition and scalar multiplication defined as :

$$(a, b) + (c, d) = (a + c, b + d) \text{ and } \alpha(a, b) = (\alpha a, b) \text{ for } \alpha \in \mathbb{R}.$$

(c) $V = \mathbb{R}$ under addition and scalar multiplication defined as :

$$a + b = a^b, \ a, b \in \mathbb{R} \text{ and } \alpha a = 0, \ \forall \ \alpha \in \mathbb{R}.$$

Q.3 If $V = \mathbb{R}^2$ under usual addition and scalar multiplication then show that following sets are subspaces of V.

(a) $H = \{(a, b) \in V \mid a = 0\}$ under usual addition and scalar multiplication defined on V.

(b) $H = \{(a, b) \in V \mid a = b\}$ under usual addition and scalar multiplication defined on V.

Q.4. (a) Explain why $H = \mathbb{R}^2$ cannot be subspace of $V = \mathbb{R}^3$ under usual addition and scalar multiplication.

(b) If $V = \mathbb{R}^3$ under usual addition and scalar multiplication and $H = \{(a, b, c) \in \mathbb{R}^3 \mid a + b + c = 1\}$. Show that H is not subspace of V.

Q.5. Which of the following sets are linearly independent or linearly dependent ?

(a) $S = \left\{ \vec{u}_1 = \begin{bmatrix} 1 \\ 0 \\ 0 \end{bmatrix}, \vec{u}_2 = \begin{bmatrix} 1 \\ 1 \\ 0 \end{bmatrix}, \vec{u}_3 = \begin{bmatrix} 1 \\ 1 \\ 1 \end{bmatrix} \right\}$ of $\mathbb{R}^3$

(b) $S = \left\{ \vec{u}_1 = \begin{bmatrix} 1 \\ 2 \\ 3 \end{bmatrix}, \vec{u}_2 = \begin{bmatrix} 4 \\ 5 \\ 6 \end{bmatrix}, \vec{u}_3 = \begin{bmatrix} 2 \\ 1 \\ 0 \end{bmatrix} \right\}$ of $\mathbb{R}^3$

(c) $S = \left\{ \vec{u}_1 = \begin{bmatrix} 1 \\ 2 \end{bmatrix}, \vec{u}_2 = \begin{bmatrix} 3 \\ 4 \end{bmatrix}, \vec{u}_3 = \begin{bmatrix} 5 \\ 6 \end{bmatrix} \right\}$ of $\mathbb{R}^3$

(d) $S = \left\{ \vec{u}_1 = \begin{bmatrix} 1 \\ 2 \\ 3 \end{bmatrix}, \vec{u}_2 = \begin{bmatrix} 4 \\ 5 \\ 6 \end{bmatrix}, \vec{u}_3 = \begin{bmatrix} 7 \\ 8 \\ 9 \end{bmatrix} \right\}$ of $\mathbb{R}^3$

(e) Any set S in $\mathbb{R}^n$ which contain zero vector.

Q.6. (a) Write the vector $v = (1, -2, 5)$ as a linear combination of the vectors $\vec{e}_1 = (1, 1, 1)$, $\vec{e}_2 = (1, 2, 3)$ and $\vec{e}_3 = (2, -1, 1)$.

(b) Find the value of α for which the vector $\vec{u} = (1, -2, \alpha)$ in $\mathbb{R}^3$ be a linear combination of the vectors $\vec{v} = (3, 0, -2)$ and $\vec{w} = (2, -1, -5)$.

(c) Find the condition on a, b and c so that $(a, b, c) \in \mathbb{R}^3$ is in S. Where, $S = \text{Span} \{\vec{u}, \vec{v}, \vec{w}\}$, $\vec{u} = (2, 1, 0)$, $\vec{v} = (1, -1, 2)$ and $\vec{w} = (0, 3, -4)$.

Q.7. Find the rank of following matrix :

(a) $A = \begin{bmatrix} 1 & 2 & 3 \\ 4 & 5 & 6 \\ 7 & 8 & 9 \end{bmatrix}$

(b) $A = \begin{bmatrix} 1 & 1 & 1 \\ 0 & 1 & 1 \\ 0 & 0 & 1 \end{bmatrix}$

(c) $A = \begin{bmatrix} 1 & 2 & -3 & 0 \\ 2 & 4 & -2 & 2 \\ 3 & 6 & -4 & 3 \end{bmatrix}$

(d) $A = \begin{bmatrix} 2 & 3 & 4 & 5 & 6 \\ 0 & 0 & 3 & 2 & 5 \\ 0 & 0 & 0 & 0 & 2 \end{bmatrix}$

Q.8. Find the co-ordinate vector of $\vec{v}$ relative to the following basis for $\mathbb{R}^3$, where $\vec{v} = (a, b, c) \in \mathbb{R}^3$.

(a) $B = \{\vec{u}_1 = (1, 0, 0),\ \vec{u}_2 = (1, 1, 0),\ \vec{u}_3 = (0, 0, 1)\}$

(b) $B = \{\vec{u}_1 = (0, 0, 1),\ \vec{u}_2 = (0, 1, 1),\ \vec{u}_3 = (1, 1, 1)\}$

(c) $B = \{\vec{u}_1 = (1, 1, 1),\ \vec{u}_2 = (1, 2, 3),\ \vec{u}_3 = (2, -1, 1)\}$

Q.9. (a) Show that in any vector space V if; $S = \{\vec{u}_1,\ \vec{u}_2,\ \vec{u}_3\}$ are independent vectors then $S_1 = \{\vec{u}_1 + \vec{u}_2,\ \vec{u}_1 - \vec{u}_2,\ \vec{u}_1 - 2\vec{u}_2 + u_3\}$ are also independent.

(b) If $S_1 \subseteq S_2$ and S_1 is linearly dependent then S_2 is also linearly dependent.

Q.10. (a) Find the basis for the null space of matrix.

$$A = \begin{bmatrix} 1 & 4 & 7 & 10 \\ 2 & 5 & 8 & 11 \\ 3 & 6 & 9 & 12 \end{bmatrix}$$

$$B = \begin{bmatrix} 2 & 4 & 6 & 8 \\ 1 & 3 & 0 & 5 \\ 1 & 1 & 6 & 3 \end{bmatrix}$$

(b) Find the basis for the column space of matrix.

$$A = \begin{bmatrix} 3 & 4 & 0 & 7 \\ 1 & -5 & 2 & -2 \\ -1 & 4 & 0 & 3 \\ 1 & -1 & 2 & 2 \end{bmatrix}$$

$$B = \begin{bmatrix} 1 & 2 & -1 & 4 \\ -2 & 1 & 7 & 2 \\ -1 & -4 & -1 & 3 \\ 3 & 2 & -7 & -1 \end{bmatrix}$$

(B) Eigenvalues and Eigenvectors

Q.1. Find the Eigenvalues and Eigenvectors of the following matrices :

$$A = \begin{bmatrix} 1 & 2 \\ 3 & 2 \end{bmatrix}, \qquad B = \begin{bmatrix} 1 & 4 \\ 2 & 3 \end{bmatrix}$$

$$C = \begin{bmatrix} 1 & -3 & 3 \\ 3 & -5 & 3 \\ 6 & -6 & 4 \end{bmatrix}, \qquad D = \begin{bmatrix} -3 & 1 & -1 \\ -7 & 5 & -1 \\ -6 & 6 & -2 \end{bmatrix}$$

$$E = \begin{bmatrix} 1 & 2 & 3 & 4 \\ 0 & 2 & 3 & 4 \\ 0 & 0 & 3 & 4 \\ 0 & 0 & 0 & 4 \end{bmatrix}, \qquad F = \begin{bmatrix} 1 & 0 & 1 \\ 0 & -1 & 2 \\ 0 & 0 & 3 \end{bmatrix}$$

Q.2. (a) Let A and B are n × n square matrices then show that AB and BA have same eigen values.

(b) Let $f(t) = 2t^2 - 5t + 6$ and $g(t) = t^3 - 2t^2 + t + 3$.

Find $f(A)$, $f(B)$, $g(A)$ and $g(B)$ where,

$$A = \begin{bmatrix} 2 & -3 \\ 5 & 1 \end{bmatrix} \text{ and } B = \begin{bmatrix} 1 & 2 \\ 0 & 3 \end{bmatrix}$$

(c) Show that matrices A and A^t have the same eigenvalues. Give an example where A and A^t have different eigenvectors.

(d) If $\vec{v}$ is an eigenvector for the matrix A corresponding to eigenvalue λ then show that $\vec{v}$ is also an eigenvector corresponding to λ^n of the matrix A^n.

Q.3. (a) Find the characteristic polynomial of the following matrices

$$A = \begin{bmatrix} 3 & -7 \\ 4 & 5 \end{bmatrix} \quad B = \begin{bmatrix} 5 & -1 \\ 8 & 3 \end{bmatrix} \quad C = \begin{bmatrix} 2 & 3 & -2 \\ 0 & 5 & 4 \\ 1 & 0 & -1 \end{bmatrix}$$

(b) Let $A = \begin{bmatrix} 1 & 1 & 0 \\ 0 & 2 & 0 \\ 0 & 0 & 1 \end{bmatrix}$ and $B = \begin{bmatrix} 2 & 0 & 0 \\ 0 & 2 & 2 \\ 0 & 0 & 1 \end{bmatrix}$

Show that A and B have different characteristic polynomial but have same minimum polynomial.

Q.4. (a) Let $A = \begin{bmatrix} 1 & 6 \\ 5 & 2 \end{bmatrix}$, $\vec{u} = \begin{bmatrix} 6 \\ -5 \end{bmatrix}$ and $\vec{v} = \begin{bmatrix} 3 \\ -2 \end{bmatrix}$. Are $\vec{u}$ and $\vec{v}$ eigenvectors of A ?

(b) Show that $\lambda = 7$ is an eigenvalue of $A = \begin{bmatrix} 1 & 6 \\ 5 & 2 \end{bmatrix}$ and find the corresponding eigenvector.

(c) Show that $\lambda = 2$ is an eigenvalue of $A = \begin{bmatrix} 4 & -1 & 6 \\ 2 & 1 & 6 \\ 2 & -1 & 8 \end{bmatrix}$. Find a basis for the corresponding eigenspace.

(d) Is $\vec{v} = \begin{bmatrix} 4 \\ -3 \\ 1 \end{bmatrix}$ is an eigenvector of $A = \begin{bmatrix} 3 & 7 & 9 \\ -4 & -5 & 1 \\ 2 & 4 & 4 \end{bmatrix}$? If yes, find the corresponding eigenvalue.

(e) Without calculation, find one eigenvalue and two linearly independent eigenvectors of $A = \begin{bmatrix} 5 & 5 & 5 \\ 5 & 5 & 5 \\ 5 & 5 & 5 \end{bmatrix}$.

(f) If l is an eigenvalue of an invertible matrix A. Show that l^{-1} is an eigenvalue of A^{-1}.

(g) Show that if A^2 is zero matrix then $l = 0$ is only eigenvalue of A.

Q.5. (a) $A = \begin{bmatrix} 0.6 & 0.3 \\ 0.4 & 0.7 \end{bmatrix}$ $\vec{v}_1 = \begin{bmatrix} 3/7 \\ 4/7 \end{bmatrix}$ $\vec{x}_n = \begin{bmatrix} 0.5 \\ 0.5 \end{bmatrix}$

(i) Find another eigenvector $\vec{v}_2$ of ($\vec{v}_1$ is an eigenvector of A).

(ii) Verify that $\vec{x}_0 = \vec{v}_1 + \alpha \vec{v}_2$.

(b) Let $A = \begin{bmatrix} 0.5 & 0.2 & 0.3 \\ 0.3 & 0.8 & 0.3 \\ 0.2 & 0 & 0.4 \end{bmatrix}$, $\vec{v}_1 = \begin{bmatrix} 0.3 \\ 0.6 \\ 0.1 \end{bmatrix}$, $\vec{v}_2 = \begin{bmatrix} 1 \\ -3 \\ 2 \end{bmatrix}$, $\vec{v}_3 = \begin{bmatrix} -1 \\ 0 \\ 1 \end{bmatrix}$.

Show that $\vec{v}_1$, $\vec{v}_2$, $\vec{v}_3$ are eigenvectors of A.

Q.6. (a) Let $A = \begin{bmatrix} 7 & 2 \\ -4 & 1 \end{bmatrix}$ Find the formula for A given that

$A = PDP^{-1}$ where $P = \begin{bmatrix} 1 & 1 \\ -1 & -2 \end{bmatrix}$ and $D = \begin{bmatrix} 5 & 0 \\ 0 & 3 \end{bmatrix}$

(b) Diagonalize the matrix $A = \begin{bmatrix} 1 & 3 & 3 \\ -3 & -5 & -3 \\ 3 & 3 & 1 \end{bmatrix}$ if possible.

(c) Diagonalize the matrix $A = \begin{bmatrix} 2 & 4 & 3 \\ -4 & -6 & -3 \\ 3 & 3 & 1 \end{bmatrix}$ if possible.

Q.7. (a) Show that if A is both diagonalizable and invertible then A^{-1} is also diagonalizable.

(b) Show that if A has n-linearly independent eigenvectors then A^T also have n-linearly independent eigenvectors.

Q.8. (a) If $A = \begin{bmatrix} a & -b \\ b & a \end{bmatrix}$ where $a, b \in \mathbb{R}$ and not both zero then show that the eigen values of A are $l = a \pm ib$ and eigenvectors $v_1 = \begin{bmatrix} 1 \\ -i \end{bmatrix}, v_2 = \begin{bmatrix} 1 \\ i \end{bmatrix}$ respectively.

(b) Let $B_1 = \{\vec{u}_1, \vec{u}_2, \vec{u}_3\}$ and $B_2 = \{\vec{v}_1, \vec{v}_2\}$ be the bases for the vector spaces V and W respectively. If $T : V \to W$ is linear transformation given by

$$T(\vec{u}_1) = 3\vec{v}_1 - 5\vec{v}_2, \; T(\vec{u}_2) = -\vec{v}_1 + 6\vec{v}_2, \; T(\vec{u}_3) = 4\vec{v}_2.$$

Find matrix for T relative to basis B_1 and B_2.

(c) Define $T : \mathbb{R}^2 \to \mathbb{R}^2$ by $T(X) = AX$, where $A = \begin{bmatrix} 7 & 2 \\ -4 & 1 \end{bmatrix}$. Find basis B for $\mathbb{R}^2$ with the property that B-matrix for T i.e. $[T]_B$ is a diagonal matrix.

(C) Inner Product of $\mathbb{R}^n$

Q.1. If $\vec{u} = \begin{bmatrix} 2 \\ -5 \\ 1 \end{bmatrix}$ and $\vec{w} = \begin{bmatrix} 3 \\ 2 \\ -3 \end{bmatrix}$. Then find

(a) $\|\vec{u}\|$ (b) $\|\vec{w}\|$

(c) $\|\vec{u} + \vec{w}\|$ (d) $\|\vec{u} - \vec{w}\|$

(e) $\vec{u} \cdot \vec{w}$ (f) $\vec{w} \cdot \vec{u}$

(g) dist $(\vec{u}, \vec{w})$

Q.2. Verify Pythagorean Theorem for following vectors :

(a) $\vec{u}_1 = \begin{bmatrix} 1 \\ 0 \\ 1 \end{bmatrix}$, $\vec{w}_1 = \begin{bmatrix} 0 \\ 1 \\ 0 \end{bmatrix}$　　　　(b) $\vec{u}_2 = \begin{bmatrix} 1 \\ 2 \\ 3 \end{bmatrix}$, $\vec{w}_2 = \begin{bmatrix} -3 \\ -2 \\ -7/3 \end{bmatrix}$

(c) $\vec{u}_3 = \begin{bmatrix} 1 \\ 0 \\ 0 \end{bmatrix}$, $\vec{w}_3 = \begin{bmatrix} 0 \\ 0 \\ 1 \end{bmatrix}$

Q.3. If $\vec{u} = \begin{bmatrix} -2 \\ 1 \end{bmatrix}$, $\vec{w} = \begin{bmatrix} 3 \\ 1 \end{bmatrix}$. Compute the following :

(a) $\dfrac{\vec{u} \cdot \vec{w}}{\vec{u} \cdot \vec{u}}$　　　　(b) $\dfrac{\vec{u} \cdot \vec{w}}{\vec{w} \cdot \vec{w}}$

(c) $\left(\dfrac{\vec{u} \cdot \vec{w}}{\vec{u} \cdot \vec{u}} \right) \vec{u}$

Q.4. Find the unit vector in the direction and opposite to the direction of given vectors :

(a) $\vec{u} = \begin{bmatrix} -30 \\ 40 \end{bmatrix}$　　　　(b) $\vec{v} = \begin{bmatrix} 8/3 \\ 2 \end{bmatrix}$

(c) $\vec{w} = \begin{bmatrix} -6 \\ 4 \\ -3 \end{bmatrix}$　　　　(d) $\vec{z} = \begin{bmatrix} 1 \\ 0 \\ 1 \end{bmatrix}$

Q.5. Determine which pairs of vectors are orthogonal :

(a) $\vec{u}_1 = \begin{bmatrix} 8 \\ -5 \end{bmatrix}$, $\vec{w}_1 = \begin{bmatrix} -2 \\ -3 \end{bmatrix}$　　　　(b) $\vec{u}_2 = \begin{bmatrix} 12 \\ 3 \\ 5 \end{bmatrix}$, $\vec{w}_2 = \begin{bmatrix} 2 \\ -3 \\ -3 \end{bmatrix}$

(c) $\vec{u}_3 = \begin{bmatrix} -3 \\ -2 \\ 5 \\ 0 \end{bmatrix}$, $\vec{w}_3 = \begin{bmatrix} 4 \\ -1 \\ 2 \\ -6 \end{bmatrix}$　　　　(d) $\vec{u}_4 = \begin{bmatrix} 1 \\ 2 \\ 3 \\ 4 \end{bmatrix}$, $\vec{w}_4 = \begin{bmatrix} -4 \\ -3 \\ -2 \\ -1 \end{bmatrix}$

Q.6. If $\vec{u}$ and $\vec{v}$ are any vectors in $\mathbb{R}^n$. Then show that :

(a) $\vec{u} \cdot \vec{v} - \vec{v} \cdot \vec{u} = 0$

(b) If $\|\vec{u}\|^2 + \|\vec{v}\|^2 = \|\vec{u} + \vec{v}\|^2$, then $\vec{u}$ and $\vec{v}$ are orthogonal.

(c) $\|\vec{u} + \vec{v}\|^2 + \|\vec{u} - \vec{v}\|^2 = 2(\|\vec{u}\|^2 + \|\vec{v}\|^2)$

Q.7. (a) Express $\vec{y}_1 = \begin{bmatrix} 6 \\ 1 \\ -8 \end{bmatrix}$ as a linear combination of following set of vectors.

$$S_1 = \left\{ \vec{u}_1 = \begin{bmatrix} 3 \\ 1 \\ 1 \end{bmatrix}, \vec{u}_2 = \begin{bmatrix} -1 \\ 2 \\ 1 \end{bmatrix}, \vec{u}_3 = \begin{bmatrix} -1/2 \\ -2 \\ 7/2 \end{bmatrix} \right\}$$

(b) Express $\vec{y}_2 = \begin{bmatrix} 8 \\ -4 \\ -3 \end{bmatrix}$ as a linear combination of following set of vectors.

$$S_2 = \left\{ \vec{u}_1 = \begin{bmatrix} 1 \\ 0 \\ 0 \end{bmatrix}, \vec{u}_2 = \begin{bmatrix} 0 \\ 1 \\ 0 \end{bmatrix}, \vec{u}_3 = \begin{bmatrix} 1 \\ 1 \\ 0 \end{bmatrix} \right\}$$

(c) Express $\vec{y}_3 = \begin{bmatrix} 9 \\ -7 \end{bmatrix}$ as a linear combination of following set of vectors.

$$S_3 = \left\{ \vec{w}_1 = \begin{bmatrix} 2 \\ -3 \end{bmatrix}, \vec{w}_2 = \begin{bmatrix} 6 \\ 4 \end{bmatrix} \right\}$$

Q.8. (a) Compute orthogonal projection of $\vec{y}_1 = \begin{bmatrix} 7 \\ 6 \end{bmatrix}$ onto $\vec{u}_1 = \begin{bmatrix} 4 \\ 2 \end{bmatrix}$.

(b) Compute orthogonal projection of $\vec{y}_2 = \begin{bmatrix} 1 \\ 7 \end{bmatrix}$ onto $\vec{u}_2 = \begin{bmatrix} -4 \\ 2 \end{bmatrix}$.

(c) Compute orthogonal projection of $\vec{y}_3 = \begin{bmatrix} -1 \\ 3 \end{bmatrix}$ onto $\vec{u}_3 = \begin{bmatrix} 1 \\ -1 \end{bmatrix}$.

Q.9. Determine which of the following sets are orthonormal ?

(a) $S_1 = \left\{ \begin{bmatrix} 1/3 \\ 1/3 \\ 1/3 \end{bmatrix}, \begin{bmatrix} -1/5 \\ 0 \\ 1/5 \end{bmatrix} \right\}$ (b) $S_2 = \left\{ \begin{bmatrix} 0.6 \\ -0.8 \end{bmatrix}, \begin{bmatrix} 0.8 \\ 0.6 \end{bmatrix} \right\}$

(c) $S_3 = \left\{ \begin{bmatrix} -2/3 \\ 1/3 \\ 2/3 \end{bmatrix}, \begin{bmatrix} 1/3 \\ 2/3 \\ 0 \end{bmatrix} \right\}$ (d) $S_4 = \left\{ \begin{bmatrix} 2 \\ -7 \\ 1 \end{bmatrix}, \begin{bmatrix} -6 \\ -3 \\ -9 \end{bmatrix}, \begin{bmatrix} 3 \\ 1 \\ 1 \end{bmatrix} \right\}$

Q.10. (a) If $\vec{u}_1 = \begin{bmatrix} 2 \\ 5 \\ -1 \end{bmatrix}$, $\vec{u}_2 = \begin{bmatrix} -2 \\ 1 \\ 1 \end{bmatrix}$ and $W = \text{span}\{\vec{u}_1, \vec{u}_2\}$. Find

orthogonal projection of $\vec{y} = \begin{bmatrix} 1 \\ 2 \\ 3 \end{bmatrix}$ onto W.

(b) If $\vec{u}_1 = \begin{bmatrix} 1 \\ 3 \\ 5 \end{bmatrix}$, $\vec{u}_2 = \begin{bmatrix} 1 \\ 3 \\ -2 \end{bmatrix}$ and $W = \mathrm{span}\{\vec{u}_1,\ \vec{u}_2\}$. Find

orthogonal projection of $\vec{y} = \begin{bmatrix} 1 \\ 3 \\ 5 \end{bmatrix}$ onto W.

(D) Quadratic Forms and Geometry of Vector Space

Q.1. If possible diagonalize the following matrices :

(a) $A = \begin{bmatrix} 3 & -2 & 4 \\ -2 & 6 & 2 \\ 4 & 2 & 3 \end{bmatrix}$ (b) $B = \begin{bmatrix} 6 & -2 & -1 \\ -2 & 6 & -1 \\ -1 & -1 & 5 \end{bmatrix}$

(c) $C = \begin{bmatrix} 4 & -1 & -1 \\ -1 & 4 & -1 \\ -1 & -1 & 4 \end{bmatrix}$ (d) $D = \begin{bmatrix} 2 & -1 & 1 \\ -1 & 2 & -1 \\ 1 & -1 & 2 \end{bmatrix}$

Q.2. (a) Show that if A is symmetric then A^2 is symmetric.

(b) Show that if A is orthogonally diagonalizable then A^2 is also diagonalizable.

Q.3. Construct spectral decomposition of following matrices :

(a) $A = \begin{bmatrix} 7 & 2 \\ 2 & 4 \end{bmatrix}$ (b) $B = \begin{bmatrix} 1 & 2 \\ 3 & 4 \end{bmatrix}$ (c) $= \begin{bmatrix} 1 & 0 \\ 0 & 2 \end{bmatrix}$

Q.4. Determine which of the following matrices are symmetric ?

(a) $A = \begin{bmatrix} 5 & -7 \\ 3 & 5 \end{bmatrix}$ (b) $B = \begin{bmatrix} 3 & 4 \\ 2 & 2 \end{bmatrix}$ (c) $= \begin{bmatrix} -6 & 2 & 0 \\ 2 & -6 & 2 \\ 0 & 2 & -6 \end{bmatrix}$

Q.5. Find the Quadratic forms of the following matrices :

$$A = \begin{bmatrix} 1 & 2 \\ 3 & 4 \end{bmatrix},\ B = \begin{bmatrix} 3 & -2 \\ -2 & 7 \end{bmatrix},\ C = \begin{bmatrix} 5 & -1/2 & 0 \\ -1/2 & 3 & 4 \\ 0 & 4 & 2 \end{bmatrix},\ D = \begin{bmatrix} 3 & 2 & 0 \\ 2 & 2 & 1 \\ 0 & 1 & 0 \end{bmatrix}$$

Q.6. Write the matrices of the given quadratic forms :

(a) $Q(\vec{X}) = 3x_1^2 + 2x_2^2 - 5x_3^2 - 6x_1x_2 + 8x_1x_3 - 4x_2x_3$

(b) $Q(\vec{X}) = 6x_1x_2 + 4x_1x_3 - 10x_2x_3$

(c) $Q(\vec{X}) = 3x_1^2 - 2x_2^2 + 5x_3^2 + 4x_1x_2 - 6x_1x_3$

(d) $Q(\vec{X}) = 4x_3^2 - 2x_1x_2 + 4x_2x_3$

Q.7. Classify the following quadratic forms, then make change of variable $\vec{X} = P\vec{y}$ that transforms the quadratic form into no cross-product term :

(a) $Q(\vec{X}) = 4x_1^2 - 4x_1x_2 + 4x_2^2$ (b) $Q(\vec{X}) = 3x_1^2 + 4x_1x_2$

(c) $Q(\vec{X}) = 2x_1^2 + 6x_1x_2 - 6x_2^2$ (d) $Q(\vec{X}) = 2x_1^2 - 4x_1x_2 - x_2^2$

Q.8. If possible, write the given vector as an affine combination of other vectors :

(a) $\vec{y}_1 = \begin{bmatrix} 4 \\ 1 \end{bmatrix}$ and $S_1 = \left\{ \vec{u}_1 = \begin{bmatrix} 2 \\ 5 \end{bmatrix}, \vec{u}_2 = \begin{bmatrix} 1 \\ 2 \end{bmatrix}, \vec{u}_3 = \begin{bmatrix} 1 \\ 3 \end{bmatrix}, \vec{u}_4 = \begin{bmatrix} -2 \\ 2 \end{bmatrix} \right\}$

(b) $\vec{y}_2 = \begin{bmatrix} 1 \\ 2 \\ 2 \end{bmatrix}$ and $S_2 = \left\{ \vec{w}_1 = \begin{bmatrix} 4 \\ 0 \\ 3 \end{bmatrix}, \vec{w}_2 = \begin{bmatrix} 0 \\ 4 \\ 2 \end{bmatrix}, \vec{w}_3 = \begin{bmatrix} 5 \\ 2 \\ 4 \end{bmatrix}, \vec{w}_4 = \begin{bmatrix} 2 \\ 0 \\ 0 \end{bmatrix} \right\}$

Q.9. Suppose $B = \{\vec{u}_1, \vec{u}_2, \vec{u}_3\}$ is a basis for $\mathbb{R}^3$. Then show that :

(a) Span $\{\vec{u}_1 - \vec{u}_2, \ \vec{u}_1 - \vec{u}_3\}$ is a plane in $\mathbb{R}^3$.

(b) aff $\{\vec{u}_1, \vec{u}_2, \vec{u}_3\}$ is a plane through $\vec{u}_1, \vec{u}_2$ and $\vec{u}_3$ in $\mathbb{R}^3$.

Q.10. Determine which of the following sets are affinely independent :

(a) $S_1 = \left\{ \vec{u}_1 = \begin{bmatrix} 1 \\ 3 \\ 7 \end{bmatrix}, \vec{u}_2 = \begin{bmatrix} 2 \\ 7 \\ 13/2 \end{bmatrix}, \vec{u}_3 = \begin{bmatrix} 0 \\ 4 \\ 7 \end{bmatrix} \right\}$

(b) $S_2 = \left\{ \vec{u}_1 = \begin{bmatrix} 3 \\ 1 \\ 5 \end{bmatrix}, \vec{u}_2 = \begin{bmatrix} 4 \\ 3 \\ 4 \end{bmatrix}, \vec{u}_3 = \begin{bmatrix} 1 \\ 5 \\ 1 \end{bmatrix} \right\}$

Q.11. Determine whether $\vec{y}_1 = \begin{bmatrix} 0 \\ 3 \\ 3 \\ 0 \end{bmatrix}$ and $\vec{y}_2 = \begin{bmatrix} -10 \\ 5 \\ 11 \\ -4 \end{bmatrix}$ are in span S, aff S,

conV S where, $S = \left\{ \vec{u}_1 = \begin{bmatrix} 3 \\ 0 \\ 6 \\ -3 \end{bmatrix}, \vec{u}_2 = \begin{bmatrix} -6 \\ 3 \\ 3 \\ 0 \end{bmatrix}, \vec{u}_3 = \begin{bmatrix} 3 \\ 6 \\ 0 \\ 3 \end{bmatrix} \right\}$

✍ ✍ ✍

MODEL QUESTION PAPER - I
F.Y.B.Sc. Computer Science
MT - 121

Time : 2 Hours   **Total Marks : 35**

Note : 1. All questions are compulsory.
         2. Figures to the right indicate full marks.

Q.1. Attempt any five of the following :   **[5]**

(a) Justify whether, $\mathbb{R}^2$ is subspace of $\mathbb{R}^3$ or not ?

(b) If $\|\vec{u}\|^2 + \|\vec{w}\|^2 = \|\vec{u} + \vec{w}\|^2$ then show that $\vec{u}$ and $\vec{w}$ are orthogonal.

(c) If A is symmetric matrix then show that A^2 is also symmetric.

(d) Show that every subset S of $\mathbb{R}^n$ which contain zero vector is always dependent.

(e) Find the eigen values of matrix $A = \begin{bmatrix} 1 & 0 \\ 0 & 4 \end{bmatrix}$.

(f) State Pythagorean theorem for vectors in $\mathbb{R}^n$.

(g) Define basis of a vector space with an example.

Q.2. Attempt any two of the following :   **[10]**

(a) State and prove Diagonal Matrix Theorem.

(b) Find the characteristic polynomial, eigen values and their multiplicities of the matrix

$$A = \begin{bmatrix} 1 & 0 & -1 \\ 2 & 3 & -1 \\ 0 & 6 & 0 \end{bmatrix}$$

(c) Let $A = \begin{bmatrix} -8 & -2 & -9 \\ 6 & 4 & 8 \\ 4 & 0 & 4 \end{bmatrix}$ and $\vec{u} = \begin{bmatrix} 2 \\ 1 \\ -2 \end{bmatrix}$. Determine if $\vec{u}$ is in Col A.

Q.3. Attempt any two of the following :   **[10]**

(a) Diagonalize the matrix $A = \begin{bmatrix} 6 & -2 & -1 \\ -2 & 6 & -1 \\ -1 & -1 & 5 \end{bmatrix}$.

(b) (i) Find the Quadratic Form of matrix $A = \begin{bmatrix} 5 & -1/2 & 0 \\ -1/2 & 3 & 4 \\ 0 & 4 & 2 \end{bmatrix}$.

   (ii) Write the matrix of the quadratic form

$$Q(\vec{X}) = 3x_1^2 + 2x_2^2 - 5x_3^2 - 6x_1x_2 + 8x_1x_3 - 4x_2x_3$$

(c) Write $\vec{y}_1 = \begin{bmatrix} 4 \\ 1 \end{bmatrix}$ as affine combination of

$$S = \left\{ \vec{u}_1 = \begin{bmatrix} 1 \\ 2 \end{bmatrix}, \vec{u}_2 = \begin{bmatrix} 2 \\ 5 \end{bmatrix}, \vec{u}_3 = \begin{bmatrix} 1 \\ 3 \end{bmatrix}, \vec{u}_4 = \begin{bmatrix} -2 \\ 2 \end{bmatrix} \right\}$$

Q.4. Attempt any one of the following :   [10]

(a) Determine whether $\vec{y}_1 = \begin{bmatrix} 0 \\ 3 \\ 3 \\ 0 \end{bmatrix}$ are in Span S, aff S, CoV S,

where, $s = \left\{ \vec{u}_1 = \begin{bmatrix} 3 \\ 0 \\ 6 \\ -3 \end{bmatrix}, \vec{u}_2 = \begin{bmatrix} -6 \\ 3 \\ 3 \\ 0 \end{bmatrix}, \vec{u}_3 = \begin{bmatrix} 3 \\ 6 \\ 0 \\ 3 \end{bmatrix} \right\}$

(b) If A is $n \times n$ symmetric matrix then show that any two eigenvectors from a different eigenspaces are orthogonal.

(c) Find the spanning set of the null space of A where,

$$A = \begin{bmatrix} 1 & -4 & -2 & 0 & 3 & -5 \\ 0 & 0 & 1 & 0 & 0 & -1 \\ 0 & 0 & 0 & 0 & 1 & -4 \\ 0 & 0 & 0 & 0 & 0 & 0 \end{bmatrix}$$

☚ ☚ ☚

MODEL QUESTION PAPER - II
F.Y.B.Sc. Computer Science
MT - 121

Time : 2 Hours — **Total Marks : 35**

Note : 1. All questions are compulsory.
2. Figures to the right indicate full marks.

Q.1. Attempt any five of the following : [5]

(a) If $S = \{(x, y) \in \mathbb{R}^2 \mid x + y = 1\}$. Explain why S is not subspace of $\mathbb{R}^2$.

(b) If A is orthogonally diagonalizable then show that A^2 is also orthogonally diagonalizable.

(c) Find the unit vector in the direction of $\vec{u} = \begin{bmatrix} -30 \\ 40 \end{bmatrix}$.

(d) Show that every subset S of $\mathbb{R}^n$ which contain zero vector is always dependent.

(e) Verify Pythagorean theorem for $\vec{u} = \begin{bmatrix} 1 \\ 0 \\ 1 \end{bmatrix}$ and $\vec{v} = \begin{bmatrix} 0 \\ 1 \\ 0 \end{bmatrix}$.

(f) Find the distance between the vector $\vec{u} = \begin{bmatrix} 2 \\ -5 \\ 1 \end{bmatrix}$ and $\vec{w} = \begin{bmatrix} 3 \\ 2 \\ -3 \end{bmatrix}$.

(g) Define dimension of a vector space with an example.

Q.2. Attempt any two of the following : [10]

(a) Show that $\vec{u} = \cos x$, $\vec{v} = \sin 2x$, $\vec{w} = \cos 2x$ are linearly independent.

(b) Show that the eigen values of any $n \times n$ triangular matrix are the entries on its main diagonal.

(c) Find a matrix A such that the set

$$W = \left\{ \begin{bmatrix} 2s + 3t \\ r + s - 2t \\ 4r + s \\ 3r - s - t \end{bmatrix} : r, s, t \text{ real numbers} \right\} \text{ is Col A.}$$

P.3

Q.3. Attempt any two of the following :   **[10]**

(a) Diagonalize the matrix $A = \begin{bmatrix} 3 & -2 & 4 \\ -2 & 6 & 2 \\ 4 & 2 & 3 \end{bmatrix}$.

(b) (i)   Find the Quadratic Form of matrix $A = \begin{bmatrix} 3 & 2 & 0 \\ 2 & 2 & 1 \\ 0 & 1 & 0 \end{bmatrix}$.

    (ii) Write the matrix of the quadratic form

$$Q(\vec{X}) = 3x_1^2 - 2x_2^2 + 5x_3^2 + 4x_1x_2 - 6x_1x_3$$

(c) Write $\vec{y}_2 = \begin{bmatrix} 1 \\ 2 \\ 2 \end{bmatrix}$ as affine combination of

$$S = \left\{ \vec{w}_1 = \begin{bmatrix} 4 \\ 0 \\ 3 \end{bmatrix}, \vec{w}_2 = \begin{bmatrix} 0 \\ 4 \\ 2 \end{bmatrix}, \vec{w}_3 = \begin{bmatrix} 5 \\ 2 \\ 4 \end{bmatrix}, \vec{w}_4 = \begin{bmatrix} 2 \\ 0 \\ 0 \end{bmatrix} \right\}$$

Q.4. Attempt any one of the following :   **[10]**

(a) Determine whether $\vec{y}_2 = \begin{bmatrix} -10 \\ 5 \\ 11 \\ -4 \end{bmatrix}$ are in Span S, aff S, CoV S,

    where, $S = \left\{ \vec{u}_1 = \begin{bmatrix} 3 \\ 0 \\ 6 \\ -3 \end{bmatrix}, \vec{u}_2 = \begin{bmatrix} -6 \\ 3 \\ 3 \\ 0 \end{bmatrix}, \vec{u}_3 = \begin{bmatrix} 3 \\ 6 \\ 0 \\ 3 \end{bmatrix} \right\}$

(b) State and prove Spectral Theorem for symmetric matrices.

(c) State and prove spanning set theorem.

Reference Book

1. Linear Algebra and its Applications (5th Ed.) by D.C. Lay, S.R. Lay and J.J. Mac Donald.

2. Elementary Linear Algebra with Supplemental Applications by Anton.